Cognitive Studies in Literature and Performance

This series offers cognitive approaches to understanding perception, emotions, imagination, meaning-making, and the many other activities that constitute both the production and reception of literary texts and embodied performances.

More information about this series at
http://www.palgrave.com/gp/series/14903

Emelie Jonsson

The Early Evolutionary Imagination

Literature and Human Nature

palgrave
macmillan

Emelie Jonsson
UiT The Arctic University of Norway
Tromsø, Norway

Cognitive Studies in Literature and Performance
ISBN 978-3-030-82737-3 ISBN 978-3-030-82738-0 (eBook)
https://doi.org/10.1007/978-3-030-82738-0

Cover illustration: bauhaus1000/GettyImages

This Palgrave Macmillan imprint is published by the registered company Springer Nature Switzerland AG
The registered company address is: Gewerbestrasse 11, 6330 Cham, Switzerland

For the ones who raised me with art and nature,
and for the one who unified my universe.

Acknowledgments

Acknowledgments, like novels, vary by temperament. Some scholars are naturally inclined to praise the human contact that has shaped their minds during a decade-long process of composition. Others, more private, restrict themselves to lists of official research bodies. This scholar tends toward the private but recoils from formalized gratitude. I hope readers of this preliminary will be tolerant.

A few mentions must be made. Though Professor Marcus Nordlund is no longer with us, this book would not exist without him. He introduced me to the field of evolutionary literary theory and allowed me to pursue it. One of the first things he told me was that "This field is full of fascinating people who love both science and literature." I have since found his claim to be true beyond his personal example. I keep memories of our wry debates, aggravating disagreements, and moments of mutual intellectual appreciation. I still hear his voice when I write, cautioning against overreaching from the data or reducing literature to abstractions. Though I don't always listen (then as now), I cherish the voice.

Marcus told me about an intensive summer course called "Evolution, Literature, and Film" held in 2011 at Aarhus University, Denmark. That is where I met Curator's Professor Joseph Carroll, pioneer of evolutionary literary theory, and the then star Ph.D. candidate Mathias Clasen. No single encounter has been more significant for my intellectual development. Mathias has gone on to become Associate Professor at Aarhus University, leader of the Recreational Fear Lab, author of books with prestigious presses, TED-talker, and oft-cited international authority on the horror genre. Meanwhile, he has remained a steadfast friend and intellectual companion—still giving and taking editorial cuts with grace, including many administered to this book. The students he raises in Aarhus have been similarly gracious. A special recognition goes to the up-and-coming authority on villains and video games, Jens Kjeldgaard-Christiansen, whose keen editorial eye has saved me from many indignities.

My greatest debt is to Joe Carroll. His hypothesis about the adaptive function of the arts first expanded my mind toward my conclusions about Darwinian myth-making, and he has helped refine those conclusions through ten years of lively discussion. His critical eye is as judicious as it is ruthless—and it is, indeed, quite ruthless. There is no one I trust more to tell me when I'm wrong, to acknowledge when I'm right, or to let my ideas be my own. He has been a supporter beyond his contributions to this book: an anchor in the tumultuous world of academic intrigue, and a Virgil for the never-ending wanderings of literary education.

I will not linger on friends and family. Suffice to say that I'm fully aware that I'm fortunate. You are, alas, entangled in this book: My parents on the wild island; my night-and-day brothers with night-and-day families; the librarian farmer's son with oceanic compassion; the militaristic engineer with a razor-sharp literary mind; and the former zoo keeper, part-time jewelry maker, who gifted me a notebook of poetry called *Human/Nature*.

I have seen acknowledgments humorously capped by the inclusion of pets. The humor is appropriate, since the dogs and cats do not care either way, but my gratitude is rather earnest. This book would have been very different if I had not lived from childhood in direct contact with uncompromisingly non-human perspectives.

Permissions

Parts of "Chapter 4: From Adventure to Utopia" were previously published in *Evolutionary Studies in Imaginative Culture* 2 (1), 2018, as "T. H. Huxley, Arthur Conan Doyle, and the Impact of Evolution on the Human Self-Narrative"; and in *Broken Mirrors: Representations of Apocalypses and Dystopias in Popular Culture* (Routledge, 2019), as "Dystopia and Utopia After Darwin: Using Evolution to Explain Edward Bulwer Lytton's *The Coming Race*." Parts of "Chapter 6: H. G. Wells's Evolutionary Imagination" were previously published in Style. 47 (3), 2013, as "The Human Species and the Good Gripping Dreams of H. G. Wells." Parts of "Chapter 7: Joseph Conrad's Evolutionary Imagination" were previously published in *Evolutionary Perspectives on Imaginative Culture* (Springer, 2020) as "*Heart of Darkness:* Joseph Conrad's Confrontation with Amoral Nature."

Contents

CHAPTER 1

Using Evolution to Explain the Evolutionary Imagination

> THE question of questions for mankind—the problem which underlies all others, and is more deeply interesting than any other—is the ascertainment of the place which Man occupies in nature and of his relations to the universe of things. Whence our race has come; what are the limits of our power over nature, and of nature's power over us; to what goal we are tending; are the problems which present themselves anew and with undiminished interest to every man born into the world.
>
> —T. H. Huxley

Darwinian evolution did something remarkable to the human imagination. It changed the stage for our ancient habit of self-definition. Our cultural histories brim with the remnants of what T. H. Huxley called our "question of questions" (2009 [1863], 57). There is scarcely a folk tale or stage tragedy that does not address "what are the limits of our power over nature, and of nature's power over us." Every political or moral vision suggests "to what goal we are tending" (or should be). From the time of eerie Paleolithic petroglyphs, pictorial art has portrayed bestial people and anthropomorphic animals, exploring what it does and does not mean to be human. From its earliest records, fiction swarms with jackal-headed Egyptian deities, Chinese monkey kings, Greek and Norse myths about talking animals and interspecies breeding, all the way to the ancient werewolves and vampires that persist in today's popular culture. Where do

E. Jonsson, *The Early Evolutionary Imagination*,
Cognitive Studies in Literature and Performance,
https://doi.org/10.1007/978-3-030-82738-0_1

humans come from? What makes us special, how do we fit into nature, and where are we going? In answer to these questions, world religions and tribal mythologies assure us that we have been chosen by supernatural forces for some purpose, that our history was shaped by grandiose events, that our concerns have universal significance, and that the world works according to human moral laws that mete out ultimate punishments and rewards. These are all answers from within our subjective human universe. Darwin, for the first time, began to provide scientific answers that gave us glimpses of ourselves from the outside. His view of our place in nature clashed with the views of mythology and religion—and, as I will argue, with the human imagination itself.

Before Darwin, human self-definition had been largely guided by human desires. Philosophers and naturalists either thought within the established religious stories or created new stories without much to check the biases of their own human minds. Though there had been hints of species change and common descent in human thought from ancient Greek and Chinese philosophy up through Darwin's near-contemporaries Jean-Baptiste Lamarck, Erasmus Darwin, and Robert Chambers, those hints had never taken the shape of naturalistic explanation detached from a cosmic purpose of perfection or transcendence (Appleman 1970, 3–45). As Huxley said, those looking for original answers to the great human questions faced "difficulties and dangers" (2009 [1863], 57). The established answers had the double benefit of being habitual and of being inherently comfortable for the human mind. For the Victorians, Darwin's theory was a fairly late, decisive blow in the ongoing struggle between naturalistic and Christian cosmology. But the discomfort it produced ran deep. The Darwinian account of life clashes with mythology generally, because as the theory itself would come to show, it runs counter to the way our minds work.

In our time, Darwin's theory has guided a synthesis of biology, anthropology, and psychology that aims to explain our species the way we would explain any other species. To the best of our current understanding, our minds have been shaped for survival and reproduction in our particular niche. We tend to focus on the human life cycle and the threats and rewards that most commonly attend it. We pay attention to life events like births, growing up, finding a place in the sustenance system, romantic bonding, parenting, and death. We worry about survival, enjoy learning and imitation, get caught up in games of trust and deception, feel at home in communities, thrill and rage in conflict with other communities, punish bullies and cheaters, strive for status, feel the deep satisfaction of constructive effort, and take pleasure in various kinds of love and imaginative life. Though we share many of these features with other animals, they come together into a specifically human profile. That human profile is imprinted on our minds. We can imagine alternative paths of life and different time periods—even magical universes, fictional technology, or empires in outer space populated by unknown sentient species. But all such imaginative visions take their shape from the basic categories of our human life cycle. Like all imaginative stories, our cosmic origin stories build the universe around human life trajectories and social relationships.

The human profile means that Darwinian evolution fits inside our imagination like a grain of sand under an eyelid. In Darwin's origin story, we are one of millions of species that have evolved on Earth, and most species have gone extinct. Our concerns are not universally relevant, only adaptations to our particular niche. Our history was shaped by innumerable small chance events. Our future may be extinction or changes so extreme that we become unrecognizable as a species. The universe works according to amoral natural laws. These answers to the great human questions are the reverse of the answers given by mythologies and artistic visions—given, that is, by the human imagination. Despite Darwin's occasional flights of rhetoric, his human origin story came without heroic characters, anthropomorphized weather, rituals and symbols, meaningful adventures, agonistic structures arranged around human concerns, and more or less directly recommended courses of life. In its most detailed form it was abstract, counterintuitive, and difficult to grasp. Thus, it posed a problem different from all other cosmological shifts, like that from pagan religions to Christianity. Rather than being faced with a full imaginative story created by other human minds, the Victorians were faced with data gradually uncovered by human minds.

Darwin's conclusions were distressing, but they were also evocative precisely because they began to tell a story like no other. Darwinian evolutionary theory revealed facts about the natural world that still cause wonder, and principles that are still hard to keep in focus without beautification. It brought to light a delicate balance between the struggle for existence and the reality of cooperation, the details of which still puzzle biologists. It forced us to think statistically, seeing species neither as essential types nor as arbitrary categories of the human mind, but as groups of subtly different individuals united by gradations of similarity and interaction. It re-painted our family tree in a way that stretched the mind to its fullest, forcing it to envision countless life forms worked on by natural forces through deep geological time, with the few that survived still undergoing perpetual transformations. The trunk of the tree bound human beings to all other organisms. Without delicacy or flair, it uncovered our descent from microbes, fish, and tree-dwelling primates. This vision was made convincing through fossils, the anatomical organization of plants and animals and the distribution of species. Evidence converged in a compelling way from fields as diverse as geology, plant and animal husbandry, geography, and embryology. Darwin's *Origin of Species* and *Descent of Man* gathered all this evidence and shaped it into a theoretical whole. And the human imagination responded. Philosophers and theologians defended the comforting self-authored origin stories or tried to encase Darwin's grain of sand in soothing qualifications. Artists engaged evolution in allegorical battle, drew heroes and villains from evolutionary history, personified the impersonal natural forces, or took the imaginative challenge itself as their topic. Anyone remotely concerned with intellectual life had to face Darwinian evolution—whether with despair, scientific admiration, utopian gusto, sneering moral superiority, or surrealist amusement.

Evolution, which has challenged the human imagination more than any other scientific discovery, can also be used to explain the human imagination. Though the evolutionary study of human behavior is a steady research program, imaginative behavior remains an uncertain frontier populated by conflicting hypotheses. I will map out that frontier, and argue for the hypothesis that our imaginative worlds help guide our flexible behavior. According to that hypothesis, our mythologies and artistic visions imbue our concerns with universal significance so that those concerns seem valuable despite their smallness on the cosmic scale. That means that our imaginative visions necessarily distort reality to some extent. But we need to see reality clearly enough to interact with it successfully, and imaginative visions can help us do that, too. Through the arts, we can accept difficult truths and cope with the most painful clashes between personal wishes and the rest of the world. The result, I argue, is that we constantly—more or less consciously—balance our need for reliable information about the world against our need to distort the world in psychologically functional ways.

Because of the friction with our imagination, Darwinian evolution replaced illusions of an anthropocentric world with illusions of a world robbed of humanity. In my view, that is a testimony to our need for imaginative worldviews. A convincing but imaginatively inaccessible origin story made millions believe that the world had become suddenly more hostile and daily activities suddenly less coherent. Darwinian evolution provided an imaginative vantage point more completely outside the perspective of our own species than any before it, and it gave only the most rudimentary explanations of specifically human forms of sociality and mental life. The new knowledge brought order into countless puzzles in the natural world and explained previously inexplicable phenomena. At the same time, it seemed to many people to reframe human morality and kindness as illusions and reveal human culture either as absurd frills on our animal nature or as a fragile force working to restrain it. The view of humans as creatures emerging from deep geological time released creative energy, opened up the possibility of exploring strange new worlds of imagination, and produced profound existential anxiety, all at the same time.

1.1 The Case for Literature

The clash between Darwinian evolution and the human imagination involves all behaviors that can be called imaginative: political visions, philosophy, scientific rhetoric, religion, and every incarnation of the arts. I will summarize some of the main scientific, philosophical, and political lines in early evolutionary thought, but the center of my study is literature. Late Victorian authors had front row seats when Darwin unearthed his cosmological raw material, and they used novels to refine it into new mythologies. They created imaginative worlds where they and their readers could mould their personal worldviews. This early Darwinian influence can be found in everything from children's fairytales and adventure stories to political allegories, science fiction, social realism, and highly symbolic meditations on the fragility of civilization. I analyze novels from the simplest to the most complex end of this spectrum: Arthur Conan Doyle's adventure story *The Lost World*, Robert Louis Stevenson's morality tale *Dr. Jekyll and Mr. Hyde*, Edward Bulwer-Lytton's horror utopia *The Coming Race*, Jack London's psychodramas *Martin Eden* and *The Sea-Wolf*, H. G. Wells's scientific romances *The Time Machine* and *The Island of Dr. Moreau*, and Joseph Conrad's vision of cosmic self-discovery, *Heart of Darkness*. This selection exemplifies fundamentally different ways of approaching the same imaginative problem, all of which have been particularly successful in their time or ours. In order to show the influence of individual identity, I analyze the personal worldviews of the three authors in my selection whose engagement with naturalistic cosmology is most complex: Jack London, H. G. Wells, and Joseph Conrad.

Literary scholars have examined the early Darwinian influence on literature, but they have not treated it as a unique imaginative event that can be illuminated by psychology. One tradition of literary scholarship, going back to the middle of the twentieth century, has described the path of ideas from science to literature (Henkin 1963 [1940]; Stevenson 1963 [1932]). I build on the archival scholarship of this tradition, and I use many of its methods, such as cultural contextualization, biographical research, close reading, and thematic analysis. However, I seek explanations beyond the influence of one text on another. Why were the scientific ideas fascinating to authors and readers in the first place, and why were those ideas imaginatively transformed, extended, or embellished the way they were? To my knowledge, no other scholar has used an evolutionary understanding of the human mind to answer these questions. Since the late twentieth century, the questions have been answered most influentially by literary scholars who appropriate Darwin to poststructuralist theories. According to this scholarly tradition, pioneered by Gillian Beer and George Levine, Darwinism is a narrative both complicit in and resistant to undesirable social structures, which scientists and authors channel according to their ideologies. Though I share this tradition's focus on the imaginative transformation of Darwinism, I depart from its theories and methods. My argument that Darwinism is a unique imaginative challenge, and that literature is one way of meeting that challenge, is based on the assumption that human minds are more than the effects of cultural norms.

The first literary authors to incorporate Darwinian themes in their novels set about filling the naturalistic world with guiding colors. The colors get some of their vividness from universally fascinating human themes: concerns with group belonging and group stability, the double-edged power of manipulating natural forces, the primal attraction of biodiversity, the horror-tinged fascination of breaking boundaries between cognitive categories, and doubts about the validity of moral structures and social norms in the face of changing environments. Those themes were used with a variety of personal motives, and marked by the intellectual and imaginative weaves of the late Victorian period. They were employed by authors as different as Arthur Conan Doyle and Charlotte Perkins Gilman, Thomas Hardy and Charles Kingsley, Jack London and Robert Louis Stevenson, H. G. Wells and Joseph Conrad. According to temperament, authors confronted nightmares about the downfall of civilization or lost themselves in dreams of ever-growing social sympathies. Depending on inclination, they portrayed the pinnacle of human evolution as artists, warriors, scientists, philanthropists, or politicians—and the antagonists as suitable opposites. Depending on philosophical, ideological, or religious affiliations, they struggled to accept natural selection or complacently reshaped it to make their visions viable. They were nonetheless responding to the same scientific account of humanity, and facing the same challenges posed by that account to the human mind.

In the following chapters, I build a framework to explain how these authors shaped Darwin's theory to fit the human imagination. I give an overview of current science about the human species, including the way imagination and the arts seem to fit into our evolutionary niche and shape our perception of the world. I delineate the main ways in which early evolutionary thinkers assimilated natural selection to their imaginations in scientific, philosophical, and political writing, thus creating the theoretical atmosphere within which novelists wrote. Finally, I analyze the myths created by Victorian novelists to make the naturalistic origin story meaningful: the agonistic structures identifying protagonists and antagonists for our species, the accommodations for a spiritual dimension, the imagined heavens and hells that might arise from the struggle for existence, and the moral attributions, anthropomorphisms, and narratives of adventure or heroism that gave human color to evolutionary history.

Despite our growing knowledge, the pressure under which Darwin put our species' self-image has hardly disappeared. In order for our lives to remain coherent in our own minds, we need imagination to illuminate truth. We need to understand where we come from both historically and mythologically, so that our sense of selves both orients us to reality and motivates us to engage with it. For the Victorians, natural selection must in some ways have fitted more uncomfortably in the mind than it does in ours. Apart from the shock of novelty, the first incarnation of Darwinian evolution understandably lacked detail. Though Darwin speculated astutely on the evolution of human emotions, social bonds and conscience, his incomplete data and uncertainty about whether habits were heritable meant he could give only the most rudimentary explanations of specifically human sociality and mental life. Immense progress has been made since Darwin's day, including the formation of new fields like genetics and neuroscience, which means that modern humans see a more adequate reflection of themselves when they face evolution. But our understanding of specifically human behavior is still quite incomplete—as is suggested by current controversies about social evolution and the adaptive function of the arts.

In the evolutionary themes of literary works from the late Victorian period, we can see the beginnings of an imaginative fight we are still fighting. We still live with the fictions their authors created. We sometimes consciously depart from those fictions and sometimes unwittingly recreate them. By applying our modern understanding of human behavior to these literary attempts to make sense of the human, we can gain a better understanding both of the literature and of the human, as well as of the particular human behavior that is literature. Perhaps that will get us closer to the scientific prospect of letting truth illuminate imagination. In any case, understanding these literary responses of the past can help illuminate our current responses to an imaginative problem that has been passed down to us unsolved.

1.2 A Current View of the Human Species

A scientific model of the peculiar needs and capacities of the human species can help us understand why Darwin's discovery caught the imagination of so many literary authors, and why they negotiated it in the particular ways they did. My work as a literary scholar is not meant to test or support this model. The model simply forms the theoretical ground for my argument about the functions and effects of literature. The biological model in this chapter, together with the analysis of Victorian evolutionary ideas in the next chapter, forms the biocultural framework I apply to novels from the late nineteenth to early twentieth century.

A biocultural framework can reach the deepest explanation of any subject that fascinates literary authors and readers—romance and death, political conflicts and religious devotion, coming-of-age and artistic ambition. The early Darwinian influence on literature is particularly interesting because it often gave rise to overt explorations of the flexibility and limitations of the human niche. A combination of literal species awareness and highly sensitive imaginations struggling with a radically changing worldview is rich ground for literary analysis. I will cover the spectrum from evolution's most elemental imaginative challenges to the way particular works of verbal art make sense of individual lives in the naturalistic world. A broad view of the human species is necessary for that purpose. To understand the fictional species-definitions of the late nineteenth century, we need to use our best current knowledge of what humans are.

Incorporating evolutionary biology, anthropology, and psychology, the first section of this chapter constructs a general outline of the human niche. Scientific knowledge is in constant development, but this outline is meant to establish a set of relatively uncontroversial facts about the human survival strategy as a basis for reaching the difficult issues most relevant to this study. Some of the most difficult issues appear in the next section, on human social organization: the survival value of cooperation, and the negotiation between individual perspectives and common concerns that underlie all human interaction, including literature. The section after that gives extended attention to human imagination: how far it can currently be defined and explained scientifically, how it might fit into the human niche, and why costly artistic behaviors have persisted through our evolutionary history across all human cultures. The final section draws out the implications of this model of the human species for our understanding of literature—most especially literature concerned with our self-image as a species.

1.3 The Human Niche

From a modern scientific perspective, there is no more reason to believe that the human species is essentially distinct from the rest of the animal kingdom than there is to believe that the Earth is the center of the universe. A stress on species continuity that simply equates peculiarly human traits with similar traits in other animals is also misleading. Our capacities for language, conscious reasoning, imagination, large-scale cooperation and cumulative culture arise from simpler, more primitive capacities continuous with characteristics in other species, but they also collectively constitute a uniquely human adaptive strategy. Other species colonize environmental niches like the air, the water, or the trees. Humans colonize a cognitive, imaginative, and cultural niche. They occupy mental space that is their peculiar province, enabling them in turn to colonize various environmental niches (Chudek and Henrich 2011; Mithen 2007; Turchin 2006, 107–39). To understand our niche as well as we do those of other animals, we must consider both the ancient physical traits and behaviors we share with the primate and mammalian lineages and the specifically human combination and elaboration of those traits (A. H. Buss 1997; Foley and Gamble 2009; Hill et al. 2009; Panksepp 2011).

The human niche has been called primarily cognitive, primarily cultural and primarily cooperative (Barrett et al. 2007; Boyd and Richerson 2007; Dunbar 2007). The pooled evidence for these three main emphases suggests that complex cognition, cultural transmission, and social organization have worked together to make us what we are today (Gangestad and Simpson 2007, 233–363; Chudek and Henrich 2011; Whiten and Erdal 2012). Humans depend on the cognitive ability to reason in symbolic and causal terms; they form theories about the physical and biological world while keeping the past and numerous possible futures suspended in the imagination. They depend on the cultural ability to interact with the environment through artefacts, passing on information and internalizing norms; they retain and modify inventions from previous generations. And finally, they depend on cooperation, banding together in groups that share jointly acquired resources, defend against other groups and delegate tasks (Hill et al. 2009). These cognitive, cultural, and social capacities are the cornerstones of the human niche, allowing us to adapt to an almost unimaginably diverse array of ecologies without major physiological change.

Through our cognitive, cultural, and cooperative capacities, we exercise extraordinary control over each other, other species, and the physical world (Geary 2007). This level of control over the environment is fairly unique. Our closest primate relatives coexist in complex social groups, save each other from traps, build nests, and fish for termites using sticks, but they do not pass down recipes for safely preparing food from poisonous sources, carve out logs and skin their prey in order to make boats to reach new resources, or study the conditions of astronautical sexual intercourse in order to set up colonies on Mars. However, the capacities that allow us to control the environment cannot be understood in isolation from our evolutionary history. Through our common descent with animals, vertebrates, mammals, and primates, there are continuities of body plan and behavior, from basic systems for approach or avoidance and physical traits like internal skeletons, spines, heads and lungs, to mammalian live birth, nursing and parental attachment, and the complex coalitions, hierarchical structures, and rudimentary tool use of primates (Foley and Gamble 2009; Shubin 2008). Such marks are retained not only in gestation and automatic startle reflexes, but also in personality traits and cognitive architecture (A. H. Buss 1997; Linden 2007; Panksepp 2011).

Our extended family history provides the conditions under which we are now interacting with our environments.

Since a main feature of our niche is the capacity to adapt to very different ecologies, it follows that the specific conditions of survival and reproduction can be very different from one human population to the other. Despite such differences, human populations share much more than an abstract dependence on complex cognition, cooperation, and cultural transmission. There are hundreds of so-called human universals—behaviors or capacities that appear in all known human cultures. These include specific items like meal times, poetic speech, theories of fortune and misfortune, mood altering substances, childhood fears, dance, concepts of intentionality, crying, gossip, and verbal distinctions between internal states of mind and the external world (Brown 2010). Taken as an unstructured list, such commonalities are evocative and may enrich our cross-cultural understanding, but in order to become truly useful for our understanding of our species they need to be structured and given particular explanations (Brown 2004). Anthropologist Hillard Kaplan and his team set out a statistical structure within which to understand specific behaviors. They have identified a set of features that they call the "human adaptive complex": the core components of the human life cycle across different cultures and ecologies (Kaplan et al. 2007, 2009). Since the reproductive cycle is what allows organisms to pass on their traits, it is necessarily the core around which dispositions and behaviors evolve.

The human adaptive complex includes physiological traits, sustenance techniques, basic social organization, a developmental trajectory, and particular lifespan features. Humans have large information-storing, energy-demanding brains and digestive systems geared toward high-quality, nutrient-dense food. They use food acquisition techniques that are often cooperative and technological, requiring intense learning and many years to master. They engage in long-term reproductive unions, including male and female parental investment, with males specializing in resource acquisition and females combining resource acquisition with direct child care. They have a three-generation system where grandparents continue to support their children and grandchildren. They are aided by networks of kin and non-kin who share resources to counteract varying availability or success, thus keeping all individuals strong enough for cooperative endeavors. They go through a long period of dependence during childhood and youth in order to develop physically and neurologically, internalize social norms and learn sustenance techniques.

They have a long lifespan compared to other primates, with relatively low child mortality and lower adult mortality, meaning that adults continue to contribute resources after their reproductive age; that way, providers outnumber the dependent young (Kaplan et al. 2007, 270).

As Kaplan, Gurven and Lancaster suggest, the combination of traits and habits in the human adaptive complex makes for "a very specialized niche" (2007, 270). Despite our behavioral flexibility and rapid adaptability, we cannot choose to be born self-sufficient, reproduce by abandoning fertilized eggs in a pond, or sustain ourselves across generations in isolation from culture and cooperation with bark as a food source. To the extent that we can approximate such divergences from a basic human life cycle—as in egg and sperm donation, survivalist excursions, specialized diets carefully selected for nutritional value, or a reclusive apartment life with food delivered to the door—that freedom is itself a result of the specifically human control of the environment achieved through cumulative culture and cooperation.

Variations between human cultures are more than just noise in a common pattern. Those variations are particular configurations of the basic conditions of human life, with particular short-term and long-term consequences that can make both individual lives and collective experiences incredibly different from each other. Particular environmental pressures make populations more or less resistant to certain diseases or temperatures, while cultural norms and technologies modify the environment to which we adapt, influencing which individuals pass on more genes. Our comparatively low reliance on physical adaptation does not mean that we have ceased to change physically. In fact, the populations that went through intense cultural development in the last 10,000 years appear to have had their genetic change sped up (Cochran and Harpending 2009; Chudek and Henrich 2011). Research on differences between human cultures—including differences through gene-culture coevolution such as the development of adult lactose tolerance in herding populations and cognitive capacities adapted to certain lifestyles—is likely to keep turning up information that is valuable both for immediate medical purposes and for understanding the plasticity of the human species. But despite the significance of such differences, we are a single species, occupying a single flexible niche: all of us have more in common with all other human populations than we have with any other known species.

1.4 Sociality and Social Organization

Cooperation is a cornerstone of the human niche, underlying the cumulative culture that makes our artefacts visible both from our planet's outer atmosphere and on the surface of its moon. "Within the last two million years our ancestors surmounted the barrier between non-cumulative social learning (common in many species) and cumulative cultural evolution" which allowed us to achieve "a phenotypic repertoire more complex and fitness-enhancing than any single individual could discover by asocial learning alone" (Chudek and Henrich 2011, 218–19). Through parental investment, pair bonds, coalitions, clans, and large-scale urban societies with internalized and actively enforced norms, we survive more effectively than any lone member of our species. Social learning makes it possible for us to pass on and build on inventions, going from an "early cultural corpus" that "might have included know-how about tool-making, fire use and food preparation" to the various intricate technologies of today that make it possible for us to subsist on glacier seals, desert cacti, and urban groceries (Chudek and Henrich 2011, 219; see also Boyd and Richerson 2007). Our ability to imitate and educate others gives us immense survival benefits (Whiten and Erdal 2012). Mathematical biologist Martin Nowak (2006) sums up this human cornerstone in a way that hints at its combined power and fragility:

> Humans are the champions of cooperation: From hunter-gatherer societies to nation-states, cooperation is the decisive organizing principle of human society. No other life form on Earth is engaged in the same complex games of cooperation and defection. (1560)

The evolution of sociality continues to be hotly debated in evolutionary biology in part because adaptations for purely altruistic behavior in biological terms run counter to evolutionary theory. Organisms with dispositions to give up their fitness for the fitness of others simply do not live long enough or reproduce well enough to pass on those dispositions. But the picture is not as simple as it may seem. Biological fitness is not limited to the well-being of a single organism; it consists of an organism's total genetic contribution to the next generation, which is influenced by the success of its genetic relatives as well as its own. For humans, that means fitness involves the capacity to survive, find a mate and raise offspring to maturity in incredibly intricate social environments.

So what we normally call altruistic behavior—anything from carrying your neighbour's groceries to risking your life for a child—is not usually altruistic in a biological sense. That is, behavior that helps others at some cost to oneself does not necessarily mean an ultimate cost in passing on one's genes. Since individual humans depend on others at every stage of their life cycle, it makes evolutionary sense for us to have dispositions for prosocial acts—and even for altruistic display. It makes sense for us to have sophisticated psychological mechanisms for empathy, sympathy and fairness, identify strongly with those closest to us, internalize norms of good behavior, and punish antisocial people (Boehm 1997; Chudek and Henrich 2011; Gintis and Van Schaik 2012; Turchin 2006, 107–39). It also makes sense for us to have more or less honest ways of navigating the social environment, to have psychological mechanisms for jealousy and domination, to be suspicious about the intentions of others, and to weigh our social contributions against our own interests (Buss 2005). Current developments in evolutionary biology promise a better understanding of the neurological mechanisms and behavioral tendencies shaped by this complex game played throughout deep time.

Nowak (2006) describes five main ways for humans to increase their fitness by helping others: investing in children, siblings, or other genetic relatives (kin selection); forming relationships of mutual aid with specific people (direct reciprocity); establishing a reputation for generosity, making strangers more likely to help in return even when they have not been personally helped (indirect reciprocity); becoming part of cooperative networks within a larger group (network reciprocity); and forming part of a cooperative group that outcompetes less cooperative groups (group selection). These five sources of cooperation require fairly specific conditions to remain evolutionarily successful. If too much effort and resources are expended with too little return, either because of too wide a circle of recipients or because of too many unloving neighbors, cooperators start losing ground. In the game of human sociality, cooperative acts are openings for defection—and defectors, in turn, run the risk of being punished by clusters of cooperators (Boehm 1997). These calculations are not explicitly present in most people's minds during most social exchanges, but they are part of the evolved emotional economy balancing our tightrope performances of cooperation.

Every aspect of sociality involves dangers and costs as well as benefits. Though we have dispositions for being generous, for taking pleasure in togetherness, and for vicariously experiencing the suffering of others, we also have individual impulses toward domination that are necessarily—but never perfectly—suppressed by society (Boehm 1997). Since no two individuals have completely overlapping interests, friction occurs on every level of sociality from family spats to world wars, claiming lives in domestic settings as well as on battlefields (Buss 2005). Imitation and social learning are low-cost routes to success, but they quickly become routes to perdition if the imitated or learned behavior is maladaptive (Morgan et al. 2012; Boyd and Richerson 2007). Though it can be lethal to trust the wrong informant, reliance on teachers and traditions is necessary to prevent every generation from inventing the various wheels that help us stay healthy and safe—build shelters, procure and cook food, and fend off predators and enemies. If a human population is to survive, tendencies to trust teachers and conform to traditions must be balanced with tendencies to doubt what we're taught and to find the truth for ourselves through direct observation (Kurzban 2007; Morgan et al. 2012). This fundamental double nature of sociality—the precarious balance between its survival value and its inherent dangers—makes it both an all-pervasive practical concern and a peculiarly absorbing topic for human self-reflection and imagination.

Because of our remarkable control of the environment and the emotional prominence of sociality, it is easy to believe that social systems are the ultimate shaping forces in human lives. Especially in post-industrial Western countries, social norms and hierarchies may seem to be the primary environment for humans to navigate. That impression is a logical result of being so far removed from our systems of subsistence that public life seems to be the main arena of dangers and possibilities. However, it is a false impression. It obscures the fact that social systems are themselves adapted to the natural environment: shaped by systems for subsistence and reproduction, which are in turn constrained both by the features of the human adaptive complex and by local geography. Though adaptation to different food sources can mean differences in fairly basic aspects of human social organization—such as the distribution between men and women of power, resource acquisition, and child care—these differences are not arbitrary. They can be predicted to a certain extent through the nature of a population's subsistence system, specifically the extent to which resources can be controlled (Foley and Gamble 2009; Kaplan et al. 2009). For instance, foragers with "mobile prey and widely distributed gathered resources" have systems of relative egalitarianism and relative monogamy, because the unpredictable resources that cannot be carried on the move make it necessary to share food and inconvenient to support many wives (Kaplan et al. 2009, 3294). But a reliance on "concentrated and predictable foraging sites"—a relatively minor modification of the nomadic foraging lifestyle, not a dramatic change in overall mode of subsistence—makes for increased polygyny and bride capture, and a stratified society with slavery and unequal access to prime foraging sites (Kaplan et al. 2009, 3294). Though these patterns are not absolute determinants, and certainly not moral guidelines, their mere existence testifies to the strong constraints exercised on social organization by our species-typical niche.

Human beings are not unreflectingly shaped by their social systems. They reach outside them to other social systems and to the external world, and they also strive to understand their social systems from within. Our niche depends not only on storing information about conspecifics, but also on storing information about other animals, plants, and the workings of the physical world; together, those areas of information make up the intuitive sciences often called folk psychology, folk biology, and folk physics (Kaplan et al. 2007; for an account of the folk sciences, see Geary 2007). These areas of knowledge are crucial for constructing functional social norms. All human societies must organize themselves around child care and extract resources from the natural environment, and must therefore have some understanding of ecology and the human life cycle—and all human subsistence systems depend on technologies that require some intuitive physics. Because fundamental problems are shared across human cultures, seven clusters of preferred moral behaviors have been identified in the ethnographic records of 60 societies: helping kin, helping your group, reciprocating, being brave, deferring to superiors, dividing disputed resources, and respecting prior possession (Curry et al. 2019). Nevertheless, as anyone familiar with twentieth-century hippie culture may attest, even these statistically preferred behaviors may be culturally opposed and vigorously debated by individuals.

The power of our social dispositions is demonstrated in the currently most extensive empirical study of literary protagonists and antagonists. This study collected data from around 500 readers who responded to characters in more than 100 novels from the nineteenth century (Carroll et al. 2012). Respondents were asked whether each character was a protagonist or antagonist and what their emotional response was to the character—but also what the character's personality and motives were like. The study found that protagonists and antagonists could be separated into recognizable patterns of motives, personality traits, and emotional effects on readers. There was one central, powerful distinction: protagonists formed cooperative clusters and antagonists exemplified dominance behavior. In fact, antagonists tended to be *exclusively* motivated by a desire for dominance, with little regard even for sex (40–3). Protagonists tended to "care about friends and family, respond to romantic attractions, and become readily absorbed in cultural pursuits" (55). This is a remarkable result if one considers the vast variation of genres, styles, values, political affiliations, and authorial personalities among nineteenth-century novels. Authors as different as Jane Austen and H. G. Wells used the same patterns to make their readers respond positively to their protagonists and negatively to their antagonists. On the basis of these data, the studies' authors argue that agonistic structure mirrors the human egalitarian syndrome: the tendency of cooperators to band together and suppress dominance behavior in individuals.

Social systems are perpetually modified in subtle ways by the individuals of which they consist. Each individual, in turn, proceeds from his or her psychological economy and assumptions about the workings of the world, which are in part derived from the individual application of folk science. That arrangement has been necessary for our survival. Taking traditions and norms from one's own social system as objective knowledge would be betting too much on limited human knowledge, on the dubious goodwill of perspectives conflicting with one's own interests, and on the imperfect nature of social learning. But social systems also have immense power to shape knowledge and fabricate norms through tradition, myth, art, ideology, and religion. One of the greatest challenges for the evolutionary humanities—and for the human mind in general—is to recognize the power and complexity of human social systems while also envisioning them in their proper relation to a larger ecology.

1.5 Imagination and the Arts

Imagination is a broad concept, at once associated with the most abstract spiritualist philosophy and the most commonsensical everyday activities. Dictionary definitions of the word list dauntingly diverse phenomena: the formation of mental images and other sensory simulation, the spontaneous associative drift characteristic of daydreaming, the consideration of possible (or impossible) courses of action, the power to produce original artefacts and ideas or frame them in novel ways, and the capacity for symbolic thinking (Roth 2007, xx). The case has been building for considering all of these phenomena together. Seemingly opposed activities like the careful weighing of alternative future scenarios and the idle drift of daydreaming are united by a peculiar cognitive capacity: the ability to experience vivid stimuli not derived from the immediate environment. Until the beginning of the twenty-first century, this cognitive capacity was mostly relegated to the realm of speculation. It was considered too vague and amorphous, or simply too complex, for scientific study (Roth 2007, xxi–xxviii). Now, it is being anatomized through the combined efforts of psychology and neuroscience.

The human imagination has been illuminated even more recently and dramatically than human social dispositions. About a decade ago, the concept had "virtually disappeared from both philosophy and mainstream psychology"; six years later, it had assumed "a central role in current efforts to understand human thought" (Roth 2007, xxiii; Taylor 2013, 792). Through studies of pretend play, mental time travel, narration, and creativity, psychologists were turning definitively toward considering imagination a key to human cognition. Vivid inner worlds were no longer a domain for select artists or a phase through which children pass (Taylor 2013, 792). Developmental psychologist Marjorie Taylor (2013) defines the object of study:

> Imagination refers to the capacity to mentally transcend time, place, and/or circumstance to think about what might have been, plan and anticipate the future, create fictional worlds, and consider remote and close alternatives to actual experiences. (791)

Taylor's definition unites dreams and hypotheses, religion and memory, musical composition and dinner plans. But she does not identify an anatomical basis for it. When she surveyed the science of imagination in 2013, she saw a core challenge for the future: identifying "neural correlates of imagination" (819). This challenge was already being met by an independent line of research. Neuroscientists had for decades built data that came together into a neural network called the Default Mode Network (DMN) (Buckner et al. 2008). The functions of this network correlate with Taylor's definition of imagination.

The DMN has been called "the imagination network" and "the crown of cognition" (Kaufman and Gregoire 2015; Carroll 2020). It is a complex neural network that activates when the mind is not focused on external tasks. Anatomically, it is located at the furthest remove from externally focused systems like vision and motor control (Margulies et al. 2016). It recruits brain areas associated with autobiographical memory, counterfactual thinking, moral evaluation, perspective taking, narrative comprehension, and spontaneous thought (Abraham 2016; Andrews-Hanna et al. 2014; Buckner et al. 2008). The DMN is engaged spontaneously by people who are unoccupied, resulting in "free and active internal mentation of the past, the future," and "non-temporal aspects of the world" (Abraham 2016, 4198). It is also engaged deliberately by people who perform internally focused tasks, such as retrieving memories or speculating about some particular event in the future (Andrews-Hanna et al. 2014). It is highly active during narrative comprehension (Simony et al. 2016). In short, this network allows us to focus our minds inward, making use of memory and simulation to create elaborate scenes, ideas, narratives, and logical connections.

The DMN has the hallmarks of an adaptation. It is a complex, costly structure: "a set of interacting brain areas that are tightly functionally connected and distinct from other systems within the brain" (Buckner et al. 2008, 4–5). It develops reliably in children, activating during behaviors such as perspective taking, mental time travel, and mind-wandering (Buckner et al. 2008, 12; Taylor 2013). The activity of the DMN is widely believed to be integral to adult human thought (Abraham 2016; Andrews-Hanna et al. 2014; Buckner et al. 2008). It appears to be the motor behind our specifically human behavioral flexibility. Its core areas developed during a period in evolutionary history that is associated with increasingly complex artefacts and paintings: a globularization of the brain that occurred in *Homo sapiens* between 100,000 and 35,000 years ago (Neubauer et al. 2018). The DMN enables our minds to create simulated worlds in which we can negotiate our impulses, internalize or resist norms, evaluate choices, plan actions, and organize aspirations and fears. Our mental simulations of the world precede, direct, and motivate our actions.

Like any adaptation, the human imagination is precariously balanced between function and dysfunction. Abnormalities in the imagination network are connected to some of the most harrowing human afflictions, such as schizophrenia, depression, autism, OCD, addiction, chronic pain, and Alzheimer's disease (Andrews-Hanna et al. 2014, 15). Some of these diseases of the mind are associated with low activity and lack of integration in the DMN, but others are associated with hyperactivity and exaggerated integration in the DMN (ibid.). At the same time, high levels of activity in the DMN are associated with creativity. At least some types of constructive imagination can be measured through the personality traits of schizotypy: unusual forms of perception, associations, and cognitive organization (Nettle 2007; Taylor 2013). The highest scorers on these traits are found among people suffering from schizophrenia, but also among mentally healthy people known for their creativity, like artists and art students. High scores predict success on tests in finding new uses for familiar objects and making quick associations between different cognitive domains; high scores also predict failure on tests in making precise distinctions using pre-assigned categories (Nettle 2007). Schizotypy tests and descriptions of the DMN both support a folk understanding of what counts as *imaginative* and *unimaginative* thinking: Unimaginative thinking means processing information according to established patterns without particular emotional investments. Imaginative thinking means departing from the information that is directly in front of us to inhabit a vivid internal world. Though there are appropriate contexts for both styles of thinking, we would be a very different species if we were not capable of imaginative thinking.

Many fields of psychology now study aspects of the imagination, but few do it with the same scope and sense of real-life significance as narrative psychology. This field, which grows out of personality psychology, focuses on the way humans structure their experiences through narratives. Narrative psychologists are particularly concerned with the stories individuals tell about their lives: "In adolescence or young adulthood, we invent stories to make sense of our own lives. And these stories change us." (McAdams 2019, 2). Self-narratives were conceived as a complement to trait-based personality theory like the Five-Factor Model (Openness to Experience, Conscientiousness, Extraversion, Agreeableness, and Neuroticism). According to narrative psychologists, identity is not just the result of such traits in interaction with the environment; it is the result of an individual's self-image over time, shaped by personality traits and basic motives in interaction with the environment, and structured as a narrative with protagonists and antagonists (McAdams and McLean 2013; McAdams and Olson 2010). This concept of "narrative identity" aligns closely with the features of the DMN: "Narrative identities reconstruct the autobiographical past and anticipate the imagined future to provide the self with temporal coherence and some semblance of psychosocial unity and purpose" (McAdams 2019, 2).

Imagination is an everyday wonder on par with the wings of a bird or the gills of a fish. Human brains have structures that transport us beyond the present. When those structures go wrong, they warp our minds and shatter our sense of self, make us feel hunted or bound by invisible patterns, disconnect us from social signals, or make us feel pain without being hurt. When they go right, they allow us to mentally travel in time; envision who we want to be; set goals; explain natural phenomena; learn cultural laws, customs, and technology; prepare for physical and moral threats—and create entire virtual worlds inside our heads. The imagination network also allows us to share those worlds through every human method of communication. In other words, it allows us to produce imaginative culture: arts like music, poetry, narrative, painting, acting and sculpture, home decoration, body ornamentation, and the artistic elements of rituals and religions.

It would be difficult to talk about *imagination* in any detail without giving special attention to *imaginative culture*. The two concepts are intertwined. As Roth puts it, "culture furnishes the material for imagination, notably, but by no means exclusively in 'high-end' forms such as works of literature... Imaginative minds are in continuous interplay with culture, both augmenting and shaping its materials, and being sustained and altered by them" (2007, xxix). Reflections on the imagination have historically centered on artistic creation—sometimes to the exclusion of non-artistic forms of imagination (Roth 2007; Paul and Kaufman 2014). This focus on the arts lingers like a ghost in modern discussions of the imagination. That ghost is negotiated consciously by the relatively new field of philosophy of creativity (Paul and Kaufman 2014). It is summoned in sophisticated, highly technical ways by recent monographs about the imagination that give center stage to spontaneity and creativity (Asma 2017; Fuentes 2017). Even if the arts are not the whole story, they are a remarkable part of the story.

Though the specific sets of arts vary from culture to culture, the production of imaginative artefacts and performances is a universal human behavior (Brown 2010; Dissanayake 2000; Dutton 2009; Hill et al. 2009). Human beings have the capacity to create and experience artistic artefacts, reliably develop the habit of using that capacity, and pour immense amounts of time and effort into it (Gottschall 2012). Why we do these things is still an open question. Roth's comment above gives what to many might be the commonsensical answer: we use the arts to furnish and express our imaginative minds. However, that answer is posed as commonsensical observation, not as scientific hypothesis. Roth does not attempt to explain the function of such furnishing and expression. From an evolutionary perspective, the arts are a riddle. Have they contributed in some way to our survival and reproduction, or are they just a form of mental home decoration that has not gotten in our way?

Several hypotheses about the function of the arts have been put forth by biologists, psychologists, and evolutionists in the humanities (Carroll 2012a). These hypotheses differ in important ways from most theories about the arts that have proliferated from Aristotle through contemporary philosophical speculation. Evolutionary hypotheses attempt to locate the universal behavior of imaginative culture in the emerging scientific understanding of the human mind. Such attempts are controversial. Many scientists and humanists alike regard the project as quixotic or simply institutionally inappropriate (Carroll et al. 2017). This reluctance to approach the arts scientifically has much in common with the attitude that held back the study of the imagination as a whole during the twentieth century. Those who put forward scientific hypotheses about the arts generally believe that the sciences and the humanities can gain equally from extending science to the arts. Without understanding the human imagination, it is hard to explain the history, formal characteristics, and effects of specific arts and art works. Without understanding the history, formal characteristics, and effects of specific arts and art works, it is hard to get a full view of the human imagination.

Five main hypotheses have sought to fit artistic behavior into the evolutionary logic of the human niche. A description of these five hypotheses follows below. I adopt one of them as the broadest and most adequate account of artistic behaviors but absorb elements from all five.

The first hypothesis suggests that imaginative culture is incidental to the human mind. According to this hypothesis, artistic behavior is a byproduct of our cognitive adaptations—that is, something to which our minds are susceptible rather than something we need. Even if the imagination network has adaptive functions, the arts might be considered functionless stimulation rather than furnishing of the imagination: a "pleasure technology" created by simulating the "goal states that served biological fitness in ancient environments, such as food, sex, safety, parenthood, friendship, status, and knowledge" (Pinker 1997, 525, 524). As triggers of adaptive dispositions, the arts might be likened to recreational drugs, processed, high-calorie sweets, or masturbation (Pinker 1997, 524–525). This hypothesis about the arts has primarily been advocated by psycholinguist Steven Pinker, who admits partial adaptive functions to some art forms, like narrative fiction (1997, 521–545; 2010; see fifth hypothesis below). It does not convincingly explain the costliness, reliable development, universality, and behavioral impact of the arts in all known human cultures. Nevertheless, the hypothesis remains relevant. From an evolutionary perspective, complex, reliably developing species-wide behavior can only be explained in two ways: as byproducts or as adaptations (including so-called "exaptations": byproducts or old adaptive traits that have been put to new adaptive uses). Hence, those who express doubt that artistic behavior has adaptive benefits—for instance, geneticists Gregory Cochran and Henry Harpending, literary scholar Jonathan Kramnick, and philosopher Stephen Davies—end up in a default position of agreement with the byproduct hypothesis (Cochran and Harpending 2009, 126–27; Kramnick 2011; Davies 2012).

The second major hypothesis suggests that the arts are an adaptation for group cohesion. In that view, artistic behavior promotes common values and norms by stigmatizing antisocial behavior and enhancing the pleasures of togetherness, thus making possible our high level of cooperation and large-scale group living (Dissanayake 2000). This hypothesis accords with the basically communicative nature of the arts, as well as with the communal nature of many art forms and the artistic aspects of communal rituals. It has received some support from the empirical study of agonistic structure, which showed that the protagonists and antagonists in hundreds of novels mirror the logic of the human egalitarian syndrome: stigmatizing antagonists as single-mindedly focused on domination and portraying protagonists as prosocial and constructive (Carroll et al. 2012; for the egalitarian syndrome, see Boehm 1997). The social cohesion hypothesis has been criticized on the grounds that many successful artists promote explicitly antisocial visions or attack common values and norms (Dutton 2009).

The third major hypothesis suggests that the arts developed through sexual selection as costly display, making them a manner of cultural peacock's tail (Miller 2010; Dutton 2009). This hypothesis might explain some of art's costly extravagance, as well as the large number of love lyrics that linger on the artist's romantic qualities, express dedication to the intended, or straightforwardly beg for sex (a convincing case might for instance be made for some of John Donne's poetry: see Winkelman 2013). However, the sexual display hypothesis fails to account for crucial aspects of the arts, such as their many communal incarnations, their epistemological, moral, and political dimensions, and the fact that both prepubescent children and all-male or all-female communities engage in artistic behavior.

The fourth hypothesis suggests that the arts are an extreme mental version of other animals' play behavior (Boyd 2009, 2010). As lion cubs pounce to practice for real fights and hunts, human minds use art to learn how to recognize patterns, categorize, identify causal relationships, and envision the perspectives of others. This parallel is made credible by the fact that "the amount of play in a species correlates highly with the flexibility of its behavior"—and humans are by far the most behaviorally flexible species (Boyd 2010, 434). It is easy to see similarities between, for instance, the pretend play of children and mature forms of theater and rituals. But the idea that the arts train our cognitive functions does not provide much ground for understanding *how* they guide our flexible behavior. Does improving "our production and processing of pattern" (Boyd 2010, 434) in itself make us more capable of interacting with the world? Does heightening our "mental flexibility" (Boyd 2010, 435) help us decide what to do? Boyd, the main proponent of this hypothesis, has recently hinted that the arts train us to recognize patterns that are particularly relevant to our species (Boyd 2013, 2018). Though Boyd does not go into much detail about these "humanly relevant" patterns—and seems to envision them as primarily social—this development edges him closer to the fifth and final hypothesis about the adaptive function of the arts (Boyd 2013, 580).

The fifth hypothesis suggests that we use the arts to create virtual realities, gaining vicarious experience to guide our behavior. This hypothesis exists in two versions. One version views the arts as a form of flight simulator, or "case-based reasoning"—a means to create a database of successful actions, the same way reading about historical chess games provides chess players with tactics to win their own games (Pinker 2010, 134; Scalise Sugiyama 2006). In this view, the simulated experience gained through art is direct and realistic. That interpretation makes sense of the moralistic and informational aspects of narrative fiction, but it does not explain the massive, enduring appeal of non-representational and fantastical forms of art, such as fantasy literature and surrealist painting. The other version of the hypothesis views the simulated experience gained through art as indirect and symbolic: as a way to find meaning through experience rather than build a database from information (Carroll 2012b). According to that meaning-making hypothesis, the core function of the arts is not to provide practical advice in a digestible form—a moral sweetened by entertainment—but to engage our senses and emotions to create an experience that alters the way we see the world in our minds. That world in our minds—our imaginative worldview—is imbued with emotions and ideas that help shape our decisions and actions.

The second version of the virtual worlds hypothesis can provide a framework for studying the arts. Neither the information-oriented version of this hypothesis nor any of the first four hypotheses offers a convincing overall explanation of artistic behavior. As overall explanations, these hypotheses leave out major aspects of the arts, and some of them carry implications about the human mind that are outdated—such as the idea that the human mind is made up of specialized modules (Carroll 2012a). However, all four hypotheses contain partial truths that can be absorbed into the idea that the arts create simulated experience to help us form imaginative worldviews. In line with the group cohesion hypothesis, the arts can promote community feeling. Creating an imaginative worldview includes forming ideas of one's society and one's reasons for belonging to it—but it also includes forming ideas about oneself and one's own interests apart from society, as well as of one's family, political coalitions and any other subcultures to which one might belong, and, of course, about the individuals and groups to whom one does not belong. Like any set of capacities, the capacities for creating, appreciating, or simply acquiring imaginative artefacts can be used for sexual display. Reproduction is central to life, and therefore likely to be central to imaginative culture. For the same reason, reproduction is surrounded by imaginative culture, as seen in the romantic use of arts like music, dance, poetry, and body ornamentation. In line with the flight simulator hypothesis, one can extract information from some imaginative artefacts or performances: Aesopian moral lessons, facts about the natural world, historical details, or plausible psychological scenarios applicable to one's own life. In line with the play hypothesis, fiction seems capable of strengthening creativity. In line with the byproduct hypothesis, some imaginative products may be little more than exploitation of our cognitive biases, as in the simple wish-fulfillment of certain forms of romance or adventure stories. All of these partial functions or effects are illuminated by being subordinated to the general idea that the arts feed into the worldviews that direct our behavior.

The imaginative worldview hypothesis overlaps in obvious ways with general theories about the imagination from psychology and neuroscience. If the imagination is a set of capacities that allows us to integrate and evaluate information in simulated mental worlds, the arts are ways of elaborating that process, making it at once more vivid, concrete, and communal. To paraphrase Roth's commonsense notion, our imaginative minds constantly interact with cultural products, shaping them and being shaped in turn. When faced with the commonsense notion, few people would disagree. Yet, if posed as a scientific hypothesis, this notion has profound implications for our understanding of specifically human behavior. If the arts help give meaning to our lives—in the sense of providing structures by which we live—that is not an exclusive concern of the humanities, incidental to our evolutionary history, biology, and basic conditions of life: it is a core element of the human niche, key to the life and death decisions of our ancestors for millennia.

Much remains to be learned about imagination and the arts. From Hume to Kant to the behaviorists and Freudians, up through the cognitive psychologists, these concepts have been largely entangled in the dichotomy between rational and irrational thought (Roth 2007, xxi–xxviii). Imagination has been split into pieces to make sense of its heterogeneous qualities, and art has been explained as sinful or divine—the lowest self-indulgence or the highest gift of humanity. Present-day academics who study human behavior still tend to see imaginative culture as something outside the scope of scientific inquiry (Carroll et al. 2017). In 2007, evolutionary anthropologist Robert Foley stated that we were "a long way from a workable, let alone unified, theory of how the imagination operates" (Roth 2007, xvii). In 2013, Taylor warned that imaginative behaviors like pretend play and narration were still being studied in isolation, without "cross-fertilization" (816). Since then, a workable theory of the imagination has emerged: it can be linked to an anatomical structure, measured in real time, and traced through evolutionary history. Taylor's cross-fertilization has begun to occur: several psychologists who study imaginative behaviors now integrate—and even implement—brain imaging in their research (Jacobs and Willems 2018). But these are only beginnings. Information about the DMN, often highly technical, has not reached far into the humanities disciplines that are most dedicated to studying imaginative behavior (Asma 2017; Paul and Kaufman 2014). The science of imagination has been more or less disconnected from debates about the adaptive function of the arts.

This book is framed by the hypothesis that the arts create virtual worlds, feeding our individual imaginative worlds, helping to guide our behavior. Future research could of course overthrow the core tenets of this hypothesis. It may turn out that artistic products have no long-term effect on behavior—that they are non-functional effects of the DMN, or that different arts have unrelated functions that make it unsuitable to study them together. But for now, this hypothesis seems the most comprehensive explanation of why humans make art. The alternative to adopting specific hypotheses is to be guided by more or less implicit ideas deriving from disciplinary conventions, local impressions, or unrestrained individual speculation—all likely to give partial and inconsistent accounts of why we are an imaginative and artistic species.

1.6 Species Identity and Literature

How do the preceding sections factor into the practical work of a literary scholar? Though verbal art is a human universal (Brown 2010), it does not yet have a clear place in our understanding of the human niche. The hypothesis propounded here suggests that literature, along with its oral antecedents, is one of many ways to produce imaginative artefacts that elaborate the imaginative worldviews guiding our behavior. That position provides a useful framework within which to study the functions and effects of particular literary works—it guards, for instance, against single-minded views of literature as manipulative, emancipatory, or escapist, and it gives insight into cross-cultural functions underlying all the specific functions of literary traditions, authorships, and texts. To go beyond the general statement that literature shapes our imaginative worldviews—to form specific ideas about how the shaping happens—requires an understanding of human emotion and cognition. What is it in literature that evokes our emotions and shapes our thoughts, and how do those responses affect our imaginative inner life?

Much excellent work has been done on what it is in literature that engages our imagination, and why. The answer to *what* has been intuited by countless aesthetic theorists and artists since the time of Aristotle: we are engaged by the core concerns of the human life cycle—births, childhoods, comings-of-age, quests for belonging, sustenance, vocations, riches and renown, social norms, justices and injustices, tradition and revolution, courtships, parenthood, romantic jealousies, status rivalries, cooperation and coalitional conflicts, dangers from predators, diseases and accidents, deaths. The answer to *why* is written all over the preceding sections of this chapter, and has for decades been given by evolutionary art theorists: we are engaged by these core concerns of the human life cycle because it has paid, on average, to be attentive to them—to form ideas about them, be attracted, repulsed, and any nuanced mixture of the two, in a way that leads to successful survival and reproduction (Carroll 1995, 2012b; Dissanayake 2000; Dutton 2009).

Human emotions, like other aspects of human behavior, are flexible. People don't feel attraction and repulsion for precisely the same things or in the same ways. The map is further complicated by the formal aspects of art that appeal to us in more indirect ways, such as the primally fascinating shapes, textures, and patterns discussed by Boyd, Dissanayake, and Dutton; in the case of literature, these aspects appear both in language rhythm and in the vicarious experiences of textures, colors, tastes, and smells created by symbolic language (Dissanayake). Nonetheless, like other aspects of human behavior, human emotions are flexible within constraints. We are predisposed to learn certain things more easily than others—to fear snakes, for instance, or to love infants (Chudek and Henrich 2011; LoBue et al. 2010; Luo et al. 2011). Authors can never be sure how their readers will react to a certain theme or formal trait, and some authors may not be entirely clear about what reaction they want. However, authors can be reasonably sure that certain things will get *some type* of emotional response: rhymes or colorful symbols, snakes eating infants or lovers torn apart by warring families are like tones optimally pitched within the human hearing range. Evolutionary literary theorists have begun to analyze basic themes and patterns rooted in the human life cycle, showing the ways those themes and patterns are shaped by different cultures and individuals (Carroll et al. 2012; Clasen 2017; Cooke 2010; Jonsson 2012; Gottschall 2008; Saunders 2010, 2012).

The themes I am investigating are more abstract than infants or snakes. My subject is the reconstruction of imaginative worldviews that took place around the time of the Darwinian revolution—the shift of educated consensus from a creationist to an evolutionary origin of humanity, and the imaginative narratives responding to the new knowledge. The core question of this book, transcending that particular subject, is how and why humans mythologize nature. Why did people respond to a scientific theory by telling fanciful stories about heroes and villains? In concrete terms, how will an evolutionary framework help answer those questions?

At the most general level, I will use this chapter's account of the human mind to argue that there was an urgent need for imaginative narratives after Darwin. The late Victorian waves of existential despair did not come only from losing the Christian worldview, with its beneficent intentionalism and its moral polarity developed across millennia, but from having that old worldview replaced with a worldview that was not imaginatively meaningful. I will argue that the literary works I discuss helped authors and readers reconstruct their imaginative worldviews. That argument is based on the hypothesis that imaginative worldviews are crucial for guiding individual human behavior—that stories are neither toys detached from life nor tools invented purely for political manipulation. They have vital psychological functions.

At a more specific level, I will rely on the research about agonistic structure and self-narratives to explain literary works. Every interpretive reading in this book assumes that the author creates a value system by using protagonists and antagonists based in evolved human social dispositions. Authors make readers disapprove of certain traits by linking those traits to pure domination (universally antagonistic) and approve of other traits by linking them to self-effacing, culturally acquisitive cooperation (universally protagonistic). That is the basic machinery literary works use to create new mythologies of nature. Like all humans, authors also create self-narratives in which they cast themselves as protagonists—narratives adapted to their personalities and life experiences, informed by basic human motives, often powerfully articulated in their literary works. Such self-narratives can be shared by readers, who may respond with different degrees of liking or hostility depending on their own personalities and life experiences. I will pay particular attention to the post-Darwinian self-narratives of Jack London, H. G. Wells, and Joseph Conrad.

Finally, this chapter's account of the human species will help evaluate all mythic responses to Darwinism. What parts of human nature are left out, emphasized, relegated to spiritual realms, or split in halves? The answers to such questions will illuminate how any given author is shaping the Darwinian world into mythology. This book is fundamentally concerned with the balance between imaginative needs and naturalistic reality. The naturalistic reality sketched out in this chapter forms a backdrop to all historical analyses and literary close readings.

References

Abraham, Anna. 2016. "The Imaginative Mind." *Human Brain Mapping* 37: 4197–4211. https://doi.org/10.1002/hbm.23300.

Andrews-Hanna, Jessica R., Jonathan Smallwood, and R. Nathan Spreng. 2014. "The Default Network and Self-Generated Thought: Component Processes, Dynamic Control, and Clinical Relevance." *Annals of the New York Academy of Sciences* 1316 (1): 29–52. https://doi.org/10.1111/nyas.12360.

Appleman, Philip, ed. 1970. *Darwin: A Norton Critical Edition*. New York: Norton.

Asma, Stephen T. 2017. *The Evolution of Imagination*. Chicago: University of Chicago Press.

Barrett, H. Clark, Leda Cosmides, and John Tooby. 2007. "The Hominid Entry into the Cognitive Niche." In *The Evolution of Mind: Fundamental Questions and Controversies*, edited by Steven W. Gangestad and Jeffry A. Simpson, 241–49. New York: Guilford Press.

Boehm, Christopher. 1997. "Impact of the Human Egalitarian Syndrome on Darwinian Selection Mechanics." *The American Naturalist* 150 (S1): 100–21. https://doi.org/10.1086/286052.

Boyd, Brian. 2009. *On the Origin of Stories: Evolution, Cognition, and Fiction*. Cambridge: Harvard University Press.

———. 2010. "Art and Evolution: The Avant-Garde as Test Case: Spiegelman in *The Narrative Corpse*." In *Evolution, Literature and Film: A Reader*, edited by Brian Boyd, Joseph Carroll, and Jonathan Gottschall, 433–56. New York: Columbia University Press.

———. 2013. "Arts, Humanities, Sciences, Uses." *New Literary History* 44: 577–96. https://doi.org/10.1353/nlh.2013.0041.

———. 2018. "The Evolution of Stories: From Mimesis to Language, from Fact to Fiction." *WIREs Cogn Sci* 9: e1444. https://doi.org/10.1002/wcs.1444.

Boyd, Robert, and Peter J. Richerson. 2007. "Cultural Adaptation and Maladaptation: Of Kayaks and Commissars." In *The Evolution of Mind: Fundamental Questions and Controversies*, edited by Steven W. Gangestad and Jeffry A. Simpson, 327–32. New York: Guilford Press.

Brown, Donald E. 2004. "Human Universals, Human Nature & Human Culture." *Daedalus* 133 (4): 47–54. https://doi.org/10.1162/0011526042365645.

———. 2010. "The Universal People." In *Evolution, Literature, and Film: A Reader*, edited by Brian Boyd, Joseph Carroll, and Jonathan Gottschall, 83–96. New York: Columbia University Press.

Buckner, Randy L., Jessica R. Andrews-Hanna, and Daniel L. Schacter. 2008. "The Brain's Default Network: Anatomy, Function, and Relevance to Disease." *Annals of the New York Academy of Science* 1134: 1–38. https://doi.org/10.1196/annals.1440.011.

Buss, Arnold H. 1997. "Evolutionary Perspectives on Personality Traits." In *Handbook of Personality Psychology*, edited by Robert Hogan, John Johnson, and Stephen R. Briggs, 346–64. San Diego: Academic Press.

Buss, David M. 2005. *The Murderer Next Door: Why the Mind Is Designed to Kill*. New York: Penguin Press.

Carroll, Joseph. 1995. *Evolution and Literary Theory*. Columbia: University of Missouri Press.

———. 2012a. "The Adaptive Function of the Arts: Alternative Evolutionary Hypotheses." *Telling Stories: Literature and Evolution*, edited by Gansel, Carsten and Dirk Vanderbeke, 50–63. Berlin: De Gruyter.

———. 2012b. "The Truth About Fiction: Biological Reality and Imaginary Lives." *Style* 46 (2): 129–60.

———. 2020. "Imagination, the Brain's Default Mode Network, and Imaginative Verbal Artifacts." In *Evolutionary Perspectives on Imaginative Culture*, edited by Joseph Carroll, Mathias Clasen, and Emelie Jonsson, 31–52. Cham, Switzerland: Springer.

Carroll, Joseph, John A. Johnson, Catherine Salmon, Jens Kjeldgaard-Christiansen, Mathias Clasen, and Emelie Jonsson. 2017. "A Cross-Disciplinary Survey of Beliefs about Human Nature, Culture, and Science. *Evolutionary Studies in Imaginative Culture* 1 (1): 1–32. https://doi.org/10.26613/esic/1.1.2.

Carroll, Joseph, Jonathan Gottschall, John Johnson, and Daniel J. Kruger. 2012. *Graphing Jane Austen: The Evolutionary Basis of Literary Meaning*. Cognitive Studies in Literature and Performance. New York: Palgrave Macmillan.

Chudek, Maciej, and Joseph Henrich. 2011. "Culture-Gene Coevolution, Norm-Psychology and the Emergence of Human Prosociality." *Trends in Cognitive Science* 15 (5): 218–26. https://doi.org/10.1016/j.tics.2011.03.003.

Clasen, Mathias. 2017. *Why Horror Seduces*. Oxford; New York: Oxford University Press.

Cochran, Gregory, and Henry Harpending. 2009. *The 10,000 Year Explosion: How Civilization Accelerated Human Evolution*. New York: Basic Books.

Cooke, Brett. 2010. "Human Nature, Utopia, and Dystopia: Zamyatin's *We*." In *Evolution, Literature, and Film: A Reader*, edited by Brian Boyd, Joseph Carroll, and Jonathan Gottschall, 381–92. New York: Columbia University Press.

Curry, Oliver Scott, Daniel Austin Mullins, and Harvey Whitehouse. 2019. "Is it Good to Cooperate? Testing the Theory of Morality-as-Cooperation in 60 Societies." *Current Anthropology* 60 (1): 47–69. https://doi.org/10.1086/701478.

Davies, Stephen. 2012. *The Artful Species: Aesthetics, Art, and Evolution*. Oxford: Oxford University Press.

Dissanayake, Ellen. 2000. *Art and Intimacy: How the Arts Began*. Seattle: University of Washington Press.

Dunbar, Robin. 2007. "Evolution of the Social Brain." In *The Evolution of Mind: Fundamental Questions and Controversies*, edited by Steven W. Gangestad and Jeffry A. Simpson, 280–87. New York: Guilford Press.

Dutton, Denis. 2009. *The Art Instinct: Beauty, Pleasure, & Human Evolution*. New York: Bloomsbury Press.

Foley, Robert, and Clive Gamble. 2009. "The Ecology of Social Transitions in Human Evolution." *Philosophical Transactions of the Royal Society of London B: Biological Sciences* 364 (1533): 3267–279. https://doi.org/10.1098/rstb.2009.0136.

Fuentes, Agustin. 2017. *The Creative Spark: How Imagination Made Humans Exceptional*. New York: Dutton.

Gangestad, Steven W., and Jeffry A. Simpson, eds. 2007. *The Evolution of Mind: Fundamental Questions and Controversies*. New York: Guilford Press.

Geary, David C. 2007. "The Motivation to Control and the Evolution of General Intelligence." *The Evolution of Mind: Fundamental Questions and Controversies*, edited by Steven W. Gangestad and Jeffry A. Simpson, 305–13. New York: Guilford Press.

Gintis, Herbert, and Carel P. Van Schaik. 2012. "Zoon Politicon: The Evolutionary Roots of Human Sociopolitical Systems." *Cultural Evolution*, edited by. Peter J. Richerson and M. H. Christiansen, 25–44. Cambridge, MA: MIT Press.

Gottschall, Jonathan. 2008. *The Rape of Troy: Evolution, Violence, and the World of Homer*. Cambridge; New York: Cambridge University Press.

———. 2012. *The Storytelling Animal: How Stories Make Us Human*. Boston: Houghton Mifflin Harcourt.

Henkin, Leo J. 1963 [1940]. *Darwinism in the English Novel, 1860–1910: The Impact of Evolution on Victorian Fiction*. New York: Russell & Russell.

Hill, Kim, Michael Barton, and A. Magdalena Hurtado. 2009. "The Emergence of Human Uniqueness: Characters Underlying Behavioral Modernity." *Evolutionary Anthropology: Issues, News, and Reviews* 18 (5): 187–200. https://doi.org/10.1002/evan.20224.

Huxley, T.H. 2009 [1863]. *Evidence as to Man's Place in Nature*. Cambridge: Cambridge University Press.

Jacobs, Arthur M., and Roel M. Willems. 2018. "The Fictive Brain: Neurocognitive Correlates of Engagement in Literature." *Review of General Psychology* 22: 147–60. https://doi.org/10.1037/gpr0000106.

Jonsson, Emelie. 2012. "'Man Is the Measure': Forster's Evolutionary Conundrum." *Style* 46 (2): 161–76.

Kaplan, Hillard S., Michael Gurven, and Jane B. Lancaster. 2007. "Brain Evolution and the Human Adaptive Complex." In *The Evolution of Mind: Fundamental Questions and Controversies*, edited by Steven W. Gangestad and Jeffry A. Simpson, 269–79. New York: Guilford Press.

Kaplan, Hillard S., Paul L. Hooper, and Michael Gurven. 2009. "The Evolutionary and Ecological Roots of Human Social Organization." *Philosophical Transactions of the Royal Society of London B: Biological Sciences* 364 (1533): 3289–299. https://doi.org/10.1098/rstb.2009.0115.

Kaufman, Scott Barry, and Carolyn Gregoire. 2015. *Wired to Create: Unraveling the Mysteries of the Creative Mind*. London: Vermillion.

Kramnick, Jonathan. 2011. "Against Literary Darwinism." *Critical Inquiry* 37 (2): 315–47. https://doi.org/10.1086/657295.

Kurzban, Robert. 2007. "Representational Epidemiology: Skepticism and Gullibility." In *The Evolution of Mind: Fundamental Questions and Controversies*, edited by Steven W. Gangestad and Jeffry A. Simpson, 357–63. New York: Guilford Press.

Linden, David J. 2007. *The Accidental Mind*. Cambridge, MA: Belknap Press of Harvard University Press.

LoBue, Vanessa, David H. Rakison, and Judy S. DeLoache. 2010. "Threat Perception Across the Life Span: Evidence for Multiple Converging Pathways." *Current Directions in Psychological Science* 19 (6): 375–79. https://doi.org/10.1177/0963721410388801.

Luo, Li Zhu, Hong Li, and Kang Lee. 2011. "Are Children's Faces Really More Appealing Than Those of Adults? Testing the Baby Schema Hypothesis Beyond Infancy." *Journal of Experimental Child Psychology* 110 (1): 115–24. https://doi.org/10.1016/j.jecp.2011.04.002.

Margulies, Daniel S., Satrajit S. Ghosh, Alexandros Goulas, Marcel Falkiewicz, Julia M. Huntenburg, Georg Langs, Gleb Bezgin, Simon B. Eickhoff, F. Xavier Castellanos, Michael Petrides, Elizabeth Jefferies, and Jonathan Smallwood. 2016. "Situating the Default-Mode Network Along a Principal Gradient of Macroscale Cortical Organization." *PNAS* 113 (44): 12574–2579. https://doi.org/10.1073/pnas.1608282113.

McAdams, Dan P. 2019. "'First We Invented Stories, Then They Changed Us': The Evolution of Narrative Identity." *Evolutionary Studies in Imaginative Culture* 3 (1): 1–18. https://doi.org/10.26613/esic.3.1.110.

McAdams, Dan P., and Brad D. Olson. 2010. "Personality Development: Continuity and Change over the Life Course." *Annual Review of Psychology* 61: 517–42. https://doi.org/10.1146/annurev.psych.093008.100507.

McAdams, Dan P., and Kate C. McLean. 2013. "Narrative Identity." *Current Directions in Psychological Science* 22 (3): 233–38. https://doi.org/10.1177/0963721413475622.

Miller, Geoffrey F. 2010. "Arts of Seduction." In *Evolution, Literature, and Film: A Reader*, edited by Brian Boyd, Joseph Carroll, and Jonathan Gottschall, 156–74. New York: Columbia University Press.

Mithen, Steven. 2007. "Seven Steps in the Evolution of the Human Imagination." In *Imaginative Minds*, edited by Ilona Roth, 3–31. Oxford; New York: Oxford University Press.

Morgan, Ted J., Luke E. Rendell, M. Ehn, William Hoppitt, and Kevin N. Laland. 2012. "The Evolutionary Basis of Human Social Learning." *Proceedings of the Royal Society of London B: Biological Sciences* 279 (1729): 653–62. https://doi.org/10.1098/rspb.2011.1172.

Neubauer, Simon, Jean-Jacques Hublin, and Philipp Gunz. 2018. "The Evolution of Modern Human Brain Shape." *Science Advances* 4 (1): eaao5961. https://doi.org/10.1126/sciadv.aao5961.

Nettle, Daniel. 2007. "A Module for Metaphor? The Site of Imagination in the Architecture of Mind." In *Imaginative Minds*, edited by Ilona Roth, 259–74. Oxford; New York: Oxford University Press.

Nowak, Martin A. 2006. "Five Rules for the Evolution of Cooperation." *Science* 314 (5805): 1560–563. https://doi.org/10.1126/science.1133755.

Panksepp, Jaak. 2011. "Brainmind, Mirror Neurons, Empathy, and Morality: What to Believe About the Evolution of the Social Mind." *The Evolutionary Review: Art, Science, Culture* 2 (1): 38–50.

Paul, Elliot Samuel, and Scott Barry Kaufman, eds. 2014. *The Philosophy of Creativity: New Essays*. Oxford; New York: Oxford University Press.

Pinker, Steven. 1997. *How the Mind Works*. New York: Norton.

———. 2010. "Art and Adaptation." *Evolution, Literature, and Film: A Reader*, edited by Brian Boyd, Joseph Carroll, and Jonathan Gottschall, 25–35. New York: Columbia University Press.

Roth, Ilona. 2007. *Imaginative Minds*. Oxford; New York: Oxford University Press.

Saunders, Judith P. 2010. "Paternal Confidence in Hurston's 'The Gilded Six-Bits.'" In *Evolution, Literature, and Film: A Reader*, edited by Brian Boyd, Joseph Carroll, and Jonathan Gottschall, 392–408. New York: Columbia University Press.

———. 2012. "Female Mate-Guarding in Lawrence's 'Wintry Peacock': An Evolutionary Perspective." *College Literature* 39 (4): 69–83. https://doi.org/10.1353/lit.2012.0042.

Scalise Sugiyama, Michelle. 2006. "Lions and Tigers and Bears: Predators as a Folklore Universal." In *Anthropology and Social History: Heuristics in the Study of Literature*, edited by H. Friedrich, F. Jannidis, U. Klein, K. Mellman, S. Metzger, and M. Willems, 319–31. Paderborn: Mentis.

Simony, Erez, Christopher J. Honey, Janice Chen, Olga Lositsky, Yaara Yeshurun, Ami Wiesel, and Uri Hasson. 2016. "Dynamic Reconfiguration of the Default

Mode Network During Narrative Comprehension." *Nature Communications* 7: 12141. https://doi.org/10.1038/ncomms12141.

Shubin, Neil. 2008. *Your Inner Fish: A Journey into the 3.5-Billion-Year History of the Human Body*. New York: Pantheon Books.

Stevenson, Lionel. 1963 [1932]. *Darwin Among the Poets*. New York: Russell & Russell.

Taylor, Marjorie. 2013. "Imagination." In *The Oxford Handbook of Developmental Psychology*, edited by Philip David Zelazo, 791–831. New York: Oxford University Press.

Turchin, Peter. 2006. *War and Peace and War: The Life Cycles of Imperial Nations*. New York: Pi Press.

Whiten, Andrew, and David Erdal. 2012. "The Human Socio-Cognitive Niche and Its Evolutionary Origins." *Philosophical Transactions of the Royal Society of London B: Biological Sciences* 367 (1599): 2119–129. https://doi.org/10.1098/rstb.2012.0114.

Winkelman, Michael A. 2013. *A Cognitive Approach to John Donne's Songs and Sonnets*. New York: Palgrave Macmillan.

CHAPTER 2

Myth-Making in Early Evolutionary Thought

> Is the philanthropist or the saint to give up his endeavours to lead a noble life, because the simplest study of man's nature reveals, at its foundations, all the selfish passions and fierce appetites of the merest quadruped? Is mother-love vile because a hen shows it, or fidelity base because dogs possess it?
>
> —T. H. Huxley, *Evidence as to Man's Place in Nature*

> The birth both of the species and of the individual are equally parts of that grand sequence of events, which our minds refuse to accept as the result of blind chance.
>
> —Charles Darwin, *The Descent of Man*

To understand how evolutionary thought interacted with literature in the late Victorian period, one needs to understand how evolution was processed imaginatively by the scientists, philosophers, and social reformers who disseminated it. The history of evolutionary thinking is fraught with imaginative distortions—and possibly even more fraught with imaginative accusations of distortions (Bowler 1989; Ghiselin 2003; Hawkins 1998; Mayr 1991; Richards 2009; Young 2007). Every prominent evolutionary thinker of the Victorian period has at some point been branded a distorter of evolution: from Herbert Spencer and Ernst Haeckel to T. H. Huxley and Charles Darwin himself. Those distortions,

E. Jonsson, *The Early Evolutionary Imagination*, Cognitive Studies in Literature and Performance, https://doi.org/10.1007/978-3-030-82738-0_2

and the acrimonious debates surrounding them, are part of the inherent friction between Darwinian evolution and the human imagination.

This chapter is predominantly about *Darwinian* evolution. Darwin was unlike any other evolutionary thinker before him—including Jean-Baptiste Lamarck, Erasmus Darwin, and Robert Chambers—in that he deliberately distanced himself from supernatural explanations like forces of complexification, divine guidance, or spontaneous generation. Other theories had suggested the ideas of common descent and species transmutation, which are on their own challenging for the human mind. However, Darwin's theory caused a uniquely urgent need for new mythologies, because it went further outside the biases of the human mind than any other theory about the origin of species.

The most original and important feature of Darwinian evolution is natural selection, the mechanism that remains the guiding principle of modern evolutionary theory (Ghiselin 2003; Young 2007). Darwin convinced people of the basic idea that species have evolved from common ancestors. But providing a plausible natural mechanism was an indispensable part of his case. Before Darwin published his theory, his staunchest future supporter, T. H. Huxley, criticized Robert Chambers's *Vestiges of the Natural History of Creation* for failing to provide any such mechanism (Huxley 1854). Though Chambers' book was a bestseller, its specious methodology was easily dismissed by scientists. The public vivisection of his book may well have added to Darwin's determination to be as scientifically stringent as possible. As he wrote to Huxley, he was "almost as unorthodox about species" as Chambers, but hoped to be "not *quite* so unphilosophical" (Darwin 1854).

Natural selection is a simple idea with vast implications. In every generation, more organisms are born than can survive and reproduce; those organisms whose traits give them some advantage over others tend to survive and pass on their traits to the next generation. Over enough time, environmental change and the natural variations between individuals can produce populations that differ so much from each other that they become distinct species. A great number of variables affect this mechanism: the overall balance of ecosystems, natural catastrophes, climate change, geographic isolation, mutation, predation, ranges of food sources, types of reproductive cycles, cooperation, parasitism and symbiosis, competition in sexually reproducing species within and between sexes—and, for humans, a significant cultural heritage that allows us to cooperate and adapt rapidly to varying climates. Nevertheless, the basic rule remains: heritable traits are differentially reproduced because of environmental pressures. This rule could hardly be less suited to remain vivid in the human imagination. It is an abstract, constant, unpleasant reality. At the core of life is a principle of death: every organism lives and reproduces at the cost of other organisms. But because natural selection is a force of change so close to human life, it is also optimally suited for being reimagined—made simpler, less disturbing, more spectacular, and more in line with our own interests.

Darwin pursued the full implications of natural selection. He worked for decades with a research program that combined original theory, experiments, extensive reading, and scientific correspondence. When he finally published *On the Origin of Species* in 1859, he supported his theory through a wide range of evidence: from paleontology, anatomy, geology, embryology, animal husbandry, and the geographical distribution of species. He did not linger on the human species in this first full exposition of his theory, but he did not exclude it either. If, as his evidence suggested, all species were related and derived from an extremely simple common ancestor, every complex trait in every organism must be the result of environmental pressures working across generations: not only obviously physical traits like limbs and eyes, but also behavior, instincts, and emotions. That generalization holds for human beings as for all other organism. Darwin acknowledged this fact in the last chapter of the *Origin:*

> In the distant future I see open fields for far more important researches. Psychology will be based on a new foundation, that of the necessary acquirement of each mental power and capacity by gradation. Light will be thrown on the origin of man and his history. (2003 [1859], 397)

Long before this passage in the conclusion, perceptive readers would have registered that Darwin's ideas applied to the human species. He followed through on this application in *The Descent of Man* in 1871 and *The Expression of the Emotions in Man and Animals* in 1872.

Through modern evolutionary psychology and biocultural research, we are arriving at Darwin's "distant future." Current knowledge about human nature can help explain why Darwinian evolution has been so fascinating to so many, and why it has had such a troubled history. Why has it been so often opposed, misunderstood, reshaped, and appropriated for ideological purposes? The case seems to be that natural selection and human evolution grate against cognitive biases that have been useful during the evolutionary history of our own species: our tendency to suppress unpleasantness that is distant in time or space, our tendency to favor control and harmony in our natural environment, our tendency toward mind–body dualism, our tendency to structure past events into storylike sequences with morally meaningful cause and effect, our tendency to consider ourselves the center of the cosmos, and our need for an overarching sense of purpose (Boehm 1997; Carroll et al. 2012; Gottschall 2012; Griskevicius et al. 2012; Haselton and Nettle 2006; Stillweli and Baumeister 1997; Willard and Norenzayan 2013).

A human mind processing natural selection runs the risk of turning it into something more comprehensible, orderly, spectacular, and purposeful. Elevating natural selection to a cosmic principle, we are likely to translate it into human social terms: the struggle for existence becomes a struggle for good or a struggle against evil. We are likely to invent goals for the evolutionary process: ideal human existences toward which it is tending. We are likely to devise strategies for circumventing evolution's materialism and randomness, and for undoing its decentralization of the human species. These universal features of evolutionary myth-making combine with cultural ideas and values, and with personality differences. Political and philosophical systems affect which parts of natural selection we emphasize or suppress, and which ideal existences we imagine as its goal. Religious belief and personal dispositions influence which conclusions we draw from humanity's common ancestry and what images we have of the natural world.

Historians of science have given ample attention to the early debates that followed Darwin's publication. For instance, Peter J. Bowler's *Monkey Trials and Gorilla Sermons* gives a nuanced account of how Darwinism changed the interactions between science and religion, and what compromises and conflicts followed the change (2007). John Hedley Brooke's comprehensive history of science and religion contains a dense chapter on Darwinian evolution that begins with "The Challenge of Darwinism" (2014, 374–437). Mike Hawkins's *Social Darwinism in European and American Thought* takes a wide view of social schemes inspired by Darwinism, from American progressivism to Russian anarchism (1998). I build on the scholarship of such historians. This book reframes the clashes between Darwinism and particular imaginative systems—religious, political, and philosophical—as part of a larger clash between the natural world and the human imagination.

2.1 Negotiating Human Uniqueness

After the *Origin* was published, its implications for the human species quickly became the focus of debate (Bowler 2007). Today, anyone marginally familiar with Darwin has seen the caricatures where he personally demonstrates our simian ancestry. The idea of such ancestry was enough for many Victorians to consider natural selection unthinkable. In 1860, Bishop Wilberforce famously called out "Darwin's Bulldog" T. H. Huxley at an Oxford debate by asking whether his monkey relations were on his grandmother's or grandfather's side. Huxley's response, to the effect that he would prefer monkey relations to Bishop relations who wasted their intelligence obstructing science, allegedly caused hissing, cheering, and fainting in the audience. Though there are no reliable verbatim accounts of the exchange between Wilberforce and Huxley, the general account is symptomatic of Huxley's public clashes over human evolution in the 1860s. He had particularly acrimonious and extended exchanges with the distinguished biologist Richard Owen.

Wilberforce, Owen, and their followers waged a holy war against humanity's primate kinship. They were convinced that human evolution was blasphemous and morally degrading because it contradicted the Bible narrative. Though the Bible narrative has a specific cultural history starting with a Semitic herding tribe in the ancient Near East, it fulfilled a cross-cultural human need for the Victorians: it provided a symbolic self-image through which they could regulate their flexible behavioral potential. Without a distinct line between human beings and other species, this self-image appeared, for many, to dissolve, making human moral principles meaningless. If human ancestors had been apes, there was no telling what human descendants might be, what some contemporary humans might still be—or what every individual could or should strive to be. That uncertainty had implications for the security of civilization as well as for the limits of morality. It raised questions about the way humans used each other and the rest of the natural world. In the service of the holy war, therefore, not only Bishops but devout scientists were willing to use their authority to obstruct scientific debate.

In 1863, Huxley published a massive attempt to set the record straight: *Evidence as to Man's Place in Nature*. The book's core chapter is devoted to establishing that the human species is part of the primate order. Most of this chapter adopts the perspective of an imagined "Saturnian" zoologist, sidelining theology and finer sensibilities to evaluate the human specimen as if it were any animal plucked for categorization. The Saturnian presents conscientious measurements of limbs and comparisons of muscles and bones in humans, apes, and monkeys. Throughout this parade of facts, Huxley takes as polemical foil the two biggest allegedly scientific arguments for keeping Man in an order of his own: the distinctness of human feet and brains. After making his case, he states that he could now close the chapter were it written only for his scientific colleagues. Those colleagues would supposedly be satisfied with facts. He will, however, for the benefit of the general public, go on to address the "repugnance" that some of them might feel about his conclusion. What follows is a powerful defence of the dignity of the evolved human species. Huxley conjures images of tigers, foxes, poets, and incrementally raised Alps to bring our humble beginnings and our great achievements into the same light. Darwin wrote to Huxley of these passages: "my admiration is unbounded... I declare I never in my life read anything grander" (Darwin 1863). In the *Descent*, eight years later, Darwin would produce comparable grandeur when reflecting on the "distasteful" conclusion "that man is descended from some lowly organised form" (1981 [1871], 404).

When Darwin and Huxley reflected on the resistance to human evolution, they were partly aiming at countering religious prejudice and refuting the faulty logic of their opponents. They were guided by a desire for truth and a commitment to their theory. At the same time, they could hardly have been immune to the imaginative problem underlying their opponents' objections. Darwin's enthusiastic response to Huxley's "grand" paragraphs, and the impassioned language of the paragraphs themselves, suggests that they were not written only for the unscientific public. Huxley's eloquence responded to concerns he felt himself. Darwin's enthusiastic response shows that he was also sensitive to those concerns.

In the beginning of his defense of Darwinian humanity, Huxley uses satirical verve to diminish the problem of "repugnance" even as he admits its imaginative force. Then, he turns all his rhetorical power toward solving it:

> On all sides I shall hear the cry—"We are men and women, not a mere better sort of apes, a little longer in the leg, more compact in the foot, and bigger in brain than your brutal Chimpanzees and Gorillas. The power of knowledge—the conscience of good and evil—the pitiful tenderness

> of human affections, raise us out of all real fellowship with the brutes, however closely they may seem to approximate us."
>
> To this I can only reply that the exclamation would be most just and would have my own entire sympathy, if it were only relevant. But, it is not I who seek to base Man's dignity upon his great toe, or insinuate that we are lost if an Ape has a hippocampus minor. On the contrary, I have done my best to sweep away this vanity. I have endeavoured to show that no absolute structural line of demarcation, wider than that between the animals which immediately succeed us in the scale, can be drawn between the animal world and ourselves; and I may add the expression of my belief that the attempt to draw a psychical distinction is equally futile, and that even the highest faculties of feeling and of intellect begin to germinate in lower forms of life. At the same time, no one is more strongly convinced than I am of the vastness of the gulf between civilized man and the brutes; or is more certain that whether 'from' them or not, he is assuredly not 'of' them. No one is less disposed to think lightly of the present dignity, or despairingly of the future hopes, of the only consciously intelligent denizen of this world. (2009 [1863], 109–10)

This is a vividly painted self-image of our species, contrasted to the rest of the animal kingdom and contained within a vision of species continuity. People like Owen are portrayed as vain and irrelevant. They are "shallow rhetoricians" who "assume authority" to claim that there is a contradiction between our uniqueness and our continuity with the rest of nature (110). In subsequent passages, Huxley suggests that the imaginative problem with human evolution can be dismissed by "a sensible child," and that the danger of moral degradation can be averted by "healthy humanity"—the "common sense of the mass of mankind" (110–11). The problem will persist only for the "cynics and the 'righteous overmuch'" who are insensible to "the nobleness of the visible world" and "the grandeur of the place Man occupies therein" (111). In short, we should focus on the beauty of reality, trust our basic sense of what is good, and let pedants worry about the finer points of our imaginative self-image. While dismissing pedants, however, we should embrace knowledge. We should internalize the scientific understanding of our origin to help us appreciate ourselves, the way we may marvel more at the Alps if we understand geology: by adding "the force of intellectual sublimity to the mere aesthetic intuition of the uninstructed beholder" (111).

Huxley's imaginative self-image is intricately made. It replaces our divine appointment with inherent moral, aesthetic, and intellectual merit.

Our humble beginnings are turned into a source of pride because of our achievements, and the issue of future change is sidelined through hints at probable steady progress. Scientific endeavor is at the core of this self-image. The climax of Huxley's defense introduces cumulative culture as the ultimate distinguishing mark of humanity, placing us on a robust imaginative platform and giving us a touch of the divine:

> Our reverence for the nobility of manhood will not be lessened by the knowledge that Man is, in substance and in structure, one with the brutes; for, he alone possesses the marvellous endowment of intelligible and rational speech, whereby, in the secular period of his existence, he has slowly accumulated and organized the experience which is almost wholly lost with the cessation of every individual life in other animals; so that now he stands raised upon it as on a mountain top, far above the level of his humble fellows, and transfigured from his grosser nature by reflecting, here and there, a ray from the infinite source of truth. (112)

High ground can be easily defended; it can keep one's treasure of experience safe and one's monkey relations at a distance. The slow growth of knowledge raises humans to a pinnacle from which culture buds into immortality. Darwin would not have been the only one resonating to this place in nature. Yet, the place is made to look safer than Huxley himself thought it. He would later express considerable concern about our grosser nature, the power of culture to amend it, and the progress of civilization (2009 [1893]). In *Man's Place*, in order to allay the imaginative concerns of his reader and himself, Huxley appeals to a commonsensical recognition of our uniqueness while weaving in a mythological vision of us as ever-improving rulers of nature. The contrast between that vision and his later concerns reveals that the weave was not entirely successful. It is, nonetheless, one of the finest efforts of verbal craftsmanship aimed at solving the imaginative problem of human evolution, and it had a massive influence. Darwin's private endorsement of it echoes publicly through the *Descent*.

Huxley's chapter contains a grim contrast between two images of humanity, but it is not between Owen's divine creation and Huxley's noble evolved creature. It is between the final image of human nobility and the initial image of humanity presented by the "scientific Saturnians":

> Let us imagine ourselves scientific Saturnians, if you will, fairly acquainted with such animals as now inhabit the Earth, and employed in discussing

> the relations they bear to a new and singular 'erect and featherless biped,' which some enterprising traveller, overcoming the difficulties of space and gravitation, has brought from that distant planet for our inspection, well preserved, may be, in a cask of rum. We should all, at once, agree upon placing him among the mammalian vertebrates; and his lower jaw, his molars, and his brain, would leave no room for doubting the systematic position of the new genus among those mammals, whose young are nourished during gestation by means of a placenta, or what are called the 'placental mammals'. (69)

Exalted civilized man and the featherless biped may not be logically contradictory, but they are visions from radically different imaginative perspectives. They are tonal and emotional opposites. The image of civilized man accords pretty well with Enlightenment ideals of scientists unveiling the work of God; the Saturnian's image displays just another living thing horribly proportionate to the cosmos. It is no accident that Huxley starts with the featherless biped and ends with "a glorious sunset" on humanity's cultural plateau (Darwin 1863). Cosmological insignificance is not a viable imaginative resting point. It can only guide motivation if it is combined with some positive values and reasons to strive for those values. Darwin constructs similar rhetorical movements in the *Descent's* chapters on morality, and in its conclusion: first providing a dispassionate outside perspective, listing facts and drawing tough-minded scientific conclusions, then stepping into the human perspective and meditating in a comforting way on the moral and imaginative implications of those conclusions. The rhetorical high points toward the end of both texts touch with concern on the bestial traits of human beings, but alleviate the concern by dwelling on the distance between their own civilization and the lifestyle of apes, "savages," and "barbarians," hoping for comparable future progress. In these passages, Darwin and Huxley seem to confront their own visceral reactions to dethroning the human species.

The chords struck by the giants of Victorian evolutionary thought were also struck by other writers attempting to process human evolution imaginatively. Those who found themselves willing or forced to accept Darwin's theory first had to struggle to internalize its non-human perspective, then had to find ways of deriving human meaning from it. Often, these processes intermingled and shaped each other. Cruelties were licensed on the authority of cosmic laws, and moralists demanded that cosmic laws have human interests. In order to build moral frameworks on the amoral natural world, the human imagination emphasized, reorganized, and suppressed facts until the natural world inspired some rousing policy or satisfying manner of life. Much of this myth-making centered on natural selection. The nature of the mechanism shaping all of life was paramount to the shaping of human life.

2.2 Myths of Order, Cooperation, and Harmony

Darwin was presumably the first to struggle with maintaining a disinterested view of natural selection, and he was arguably the most successful. The boldness of his speculative intelligence was tempered by his methodology, his personal cautiousness, and the criticism of his correspondents. Nevertheless, he was at war with his own mind about the reality and moral implications of natural selection. Even if he did not formulate as complete a mythological vision of evolution as many of his contemporaries, he did occasionally shape his discovery to suit the human imagination. He also reflected openly on the difficulties of assimilating natural selection. This combination of awareness and rudimentary myth-making would on its own have made him a useful case study for understanding the imaginative transformation of natural selection. As it is, he is also its discoverer and the mentor of its two most successful popularizers, Huxley and Ernst Haeckel. Darwin is thus an obvious subject of analysis whether measured by inherent quality, historical significance, or popular visibility.

Before integrating natural selection within one's imaginative worldview, one needs to take it in and make it stick in the mind. Taking it in means facing the unpleasantness of competition and limited resources in nature, and also generalizing that unpleasantness as a rule that affects all of life. Darwin, who clearly had a mind suited for abstract reasoning, only made that extrapolation after years of field experience on the *HMS Beagle*, extensive scientific reading, and a final theoretical jolt from Malthus's reflections on population growth. Despite then spending two decades developing a theory based on the extrapolation, he observes in the *Origin* that he finds it difficult to hold on to the worldview implicit in the theory: "nothing is easier than to admit in words the truth of the universal struggle for life," but nothing is "more difficult—at least I found it so—than constantly to bear this conclusion in mind" (2003 [1859], 133). He does not speculate about the cause of this difficulty, but he identifies it as a serious problem for naturalists. Losing sight of the struggle for life results in a dim or distorted view of "the whole economy of nature, with every fact on distribution, rarity, abundance, extinction, and variation" (ibid.). This statement remains true despite the subsequent development of evolutionary theory.

The difficulty Darwin identified may derive in great part from our social dispositions. His own reflections on human sociality serve both for clarification of the dispositions and examples of their effects. In the *Descent*, he describes our tendency to forget the seriousness of self-preserving passions when we are safe:

> The desire to satisfy hunger, or any passion, such as vengeance, is in its nature temporary, and can for a time be fully satisfied. Nor is it easy, perhaps hardly possible, to call up with complete vividness the feeling, for instance, of hunger; nor indeed, as has often been remarked, of any suffering. The instinct of self-preservation is not felt except in the presence of danger; and many a coward has thought himself brave until he has met his enemy face to face. (1981 [1871], 90)

Darwin contrasts this transient perception of suffering with our more constant social dispositions. In his view, the pangs of conscience come from the fact that prosocial feelings seem stronger in hindsight than the suffering that made us act selfishly:

> At the moment of action, man will no doubt be apt to follow the stronger impulse; and though this may occasionally prompt him to the noblest deeds, it will far more commonly lead him to gratify his own desires at the expense of other men. But after their gratification, when past and weaker impressions are contrasted with the ever-enduring social instincts, retribution will surely come. (91)

Darwin's reflections seem prescient in light of contemporary theories about human sociality. However, intuition and awareness did not save Darwin from the distorting effect of security. Directly following the passage just quoted, he imagines a future where suffering and competing interests have been removed: man will "through long habit acquire such perfect self-command, that his desires and passions will at last instantly yield to his social sympathies, and there will no longer be a struggle between them" (91). This will mean that a hungry or vengeful man "will not think of stealing food, or of wreaking his vengeance"—unless he is an "essentially bad man," "destitute of sympathy," and even then he may control himself through a rational consideration of his punishment (91–92). Later in the *Descent*, Darwin evokes this utopian vision on a grander scale still: "As man advances in civilisation, and small tribes are united into larger communities," our improved social instincts will first extend to whole nations, then to "the men of all nations and races," then, as our sympathies become "more tender and more widely diffused," "spread through instruction and example to the young, and eventually through public opinion," they will extend to "all sentient beings" (100–1). Followed to its logical conclusion, this is a world of perfect human self-command where no one hurts either other human beings or sentient animals out of self-interest. The vision is predicated on the idea, then still current, that habits could be biologically "inherited" so that every new generation was born with the moral habits of the last, resulting in a progressive refinement of our social instincts (92).

Darwin's vision of human moral evolution exemplifies one of the most common mythological adaptations of natural selection among nineteenth century thinkers. People in different political positions—liberals like Darwin, eclectic socialists like Graham Wallas and H. G. Wells, anarcho-communists like Emile Gautier and Petr Kropotkin—espoused versions of the idea that evolution could lead to universal social harmony (Hawkins 1998, 82–184). They found different and sometimes opposing ways of accommodating natural selection to this vision. In Wells's version, as in Darwin's, human nature would be progressively improved by beneficial social institutions and moral habits, thus fitting "the round Palaeolithic man into the square hole of the civilised state" (1975 [1897], 221). In Wallas's version, trying to shape humanity through institutions would lead to "constant and cruel disillusion," because human nature changed so slowly that it might be considered fixed; harmony could be achieved only by changing the institutions so that they grated less against our natural shape (Hawkins 1998, 162, 163–68). Regardless of the means, these visions had a common end: suppressing the struggle for existence on which natural selection was predicated. To use Darwin's phrase, they relied on a future where "the whole economy of nature" would be radically different. Remaining internally consistent was easier for visionaries who, like Alfred Russel Wallace, George Bernard Shaw, and Karl Marx, held that natural selection could not explain the human mind and human social behavior (Hawkins 1998, 152–54, 161, 182–83). But such visionaries were instead faced with an external pressure: to explain why the human mind and social behavior were distinct from the rest of nature.

Unlike many others, Darwin did not form a political theory based on the evolution toward universal cooperation. In the *Descent*, the utopian blip is counteracted by dispassionate analyses of the conflicting impulses currently governing sociality in humans and other animals. His account of the human conscience is just one example of his ability to trace the evolutionary logic behind behavioral dispositions and weigh their relative strength in particular scenarios. The second edition of the *Descent*, in 1874, expands the discussion of conscience to include public personas and social norms. He stresses the importance of the "rules held sacred by the tribe" and a man's "deep regard for the good opinion of his fellows" (1981, 114, 115). Glancingly, he also discusses the importance of individual differences for sociality: "How far each man values the appreciation of others, depends on the strength of his innate or acquired feeling of sympathy; and on his own capacity for reasoning out the remote consequences of his act" (114). These intuitions about conscience measure well against the most advanced contemporary theories of human sociality. Darwin describes a biocultural interaction of species-wide and individually variable moral dispositions.

Sympathetic dispositions and responsiveness to social norms surely influenced people to imagine a world without the struggle for existence. Darwin had a considerable capacity for sympathy, visible for instance in his concern for animal suffering throughout the *Origin* and *Descent*. He likely shared this personality trait with many other visionaries of universal cooperation. Many such visionaries were politically or philosophically committed to social harmony before they knew of natural selection, and were inclined to bend the new biological science to the "sacred rules of their tribes." Underneath individual motivations, the tendency to view cooperation as an ultimate evolutionary goal gained fuel from just those human impulses it over-emphasized. It universalized a secure, prosocial mindset in a fully controlled natural environment with seemingly unlimited resources: an archetypal human utopia, seductive for many single minds but seldom sustainable across the different interests of two minds. The tendency was sufficiently recognizable and widespread by 1895 for H. G. Wells to eviscerate a book under the heading "Bio-Optimism" (years before he developed his own biologically grounded utopia). An early twentieth-century line of pacifism, descended from these optimists, was categorized by scholars as "peace biology" (Hawkins 1998, 183).

2.3 Myths of Competition, Inequality, and Individualism

As Darwin supposed, we have dispositions for self-interest as well as for self-sacrificing prosociality. Under pressure, the expansion of sympathy follows certain patterns of relatedness and familiarity: if forced to preserve a random sentient being on the one hand or ourselves, a family member or a friend on the other, the choice would not be difficult for most of us—even if the random sentient being were human. Selfish impulses are often suppressed by cultural norms as well as by social dispositions. For Darwin, with his high capacity for sympathy and Victorian ethos, preserving the self above others is the most primitive human impulse, and preserving the in-group above other groups is our most primitive social impulse. The "confinement of sympathy to the same tribe" is a major cause of "the low morality of savages, as judged by our standard" (97). This primitive social instinct makes a man willing to aid his fellows only as long as it does not "too greatly interfere with his own welfare or his own strong desires"—it "never extends to all the individuals of the same species" (85).

In the *Descent*, Darwin supposes that the social instinct is the material from which civilization conjures "the disinterested love for all living creatures, the most noble attribute of man" (1981 [1871], 105). This development has given us a tendency to aid and sympathize with "the helpless," and Darwin argues that we could not "check our sympathy, if so urged by hard reason, without deterioration in the noblest part of our nature" (168–69). Nevertheless, he concedes that the progress toward universal sympathy is slowed by hostility toward outgroups: when others are separated from us "by great differences in appearance or habits, experience unfortunately shews us how long it is before we look at them as our fellow-creatures" (101). In Darwin's case, "hard reason" produced an analysis of the superior and inferior groups of his own society, with statistics on the high mortality rates of the "helpless" to reassure his readers that civilized sympathy did not degrade our stock very much (167–84). Such calculations were given significantly more space in the writing of some other early evolutionary thinkers. More often than not, the labels of inferiority and superiority were tinged with self-interest, sympathy was confined to some particular superior group, and natural selection was recruited on the side of that group.

The early evolutionists who emphasized struggle and competition have been more visible than those who emphasized cooperation. Hawkins describes a tendency among current scholars to see Darwinian interpretations of society as inherently branded with "ideologies such as *laissez-faire* liberalism, racism or imperialism" (1998, 7). The awareness of such ideological links is not new. Toward the end of the nineteenth-century, German politician August Bebel argued against the use of Darwinism to support aristocratic hierarchies (Bebel 1910). In 1922, philosopher John Dewey noted and critiqued the way Darwinism had been recruited to support "laissez-faire social philosophy" (Dewey 1922, 297). nineteenth-century sociologist Jacques Novicow even equated social Darwinism with pro-war ideology (Hawkins 1998, 7). Of course, the people who spoke up against this tendency had their own moral perspectives, and often recruited Darwin on their side while condemning their opponents' use of him. Bebel, social democratic in orientation, pronounced that "Darwinism, like every real science, is eminently democratic" (1910, 254). Dewey put forth an intricate Darwinian worldview celebrating the arts (1958).

Darwinian evolution (like any scientific theory) has been enlisted across the political spectrum. During the late nineteenth and early twentieth century, it was used in support of "militarism and pacifism, capitalism and socialism, patriarchy and feminism, totalitarianism and anarchism" (Hawkins 1998, 8). All the same, an immense amount of scholarly effort has been expended entangling Darwinian evolution with and disentangling it from ideologies stressing competition—particularly militant racist agendas (Richards 2009, 269–77, 489–512). A partial explanation for that emphasis may be our preoccupation with the Nazi regime, and the shimmer of scientific plausibility that regime spread over a racial mythology that roused millions to endorse atrocities (Hawkins 1998, 184–248; Richards 2009, 489–512). A further reason might be the early association between natural selection and prominent laissez-faire philosopher Herbert Spencer. On a more fundamental level, one might suspect that competition exercises an inherent fascination not equalled by the fascinations of peace and cooperation.

We may be disposed to suppress abstract universal struggle, and to forget unpleasant realities that have no simple solutions, but we are eminently disposed to pay attention to simplified conflicts between human groups. Many evolutionary thinkers emphasizing competition—Spencer, Clémence-Auguste Royer, Georges Vacher de Lapouge, Joseph Le Conte—have envisioned human evolution as some such simplified conflict (Hawkins 1998, 184–215). They identify an outgroup set against an ingroup: Spencer set a general lower-class "unfit" against society's successful, Royer set inferior races against Aryans, Lapouge set dark-complexioned *Homo Alpinus* against pale and blue-eyed *Homo Europaeus*, and Le Conte set African American slaves against their white masters (Hawkins 1998, 82–103, 124–30, 191–200, 200–3). The conflict between the groups is described with apocalyptic urgency, playing on both self-preservation and a deep-rooted distaste for free-riders. The resulting narrative of small, worthy populations besieged by degenerates is archetypally compelling for those considering themselves members of the worthy population. At the same time, proclaiming yourself superior is a good way of making enemies. People arguing for the superiority of their own group and the elimination of others make it easy for others to transform them into archetypal villains—the all-evil bullies that stimulate egalitarian dispositions, making us want to band together and punish them. Such bullies, seeking nothing but power and motivated by nothing but hate, serve as the optimal symbols against which cooperative groups may band together. Therefore, the evolutionary thinkers emphasizing competition become the optimal outgroup for those stressing cooperation and social harmony. Both the competition they advocate and the competition *against* them stirs what Darwin called our most primitive social instinct.

Evolutionary thinkers, like all human beings, proceed to some degree from the "sacred rules of their tribes." Many evolutionary thinkers emphasizing cooperation used natural selection to scaffold their preexisting affiliations with some ideological group (Hawkins 1998, 151–83). Some of them, like H. G. Wells and Charlotte Perkins Gilman, gained much fuel from antagonism toward an outgroup—in Wells's case, anti-scientific people and militarists; in Gilman's case, masculine men (Hawkins 1998, 262–64, 269–70; Orwell 1941). Likewise, the evolutionary thinkers emphasizing competition used natural selection to advocate hierarchies of race and class that had been current before Darwin's discovery (Hawkins 1998, 184–215; Richards 2009, 489–512). They were far from agreeing how the outgroups should be treated. For instance, Spencer and Royer both opposed war, believing it to be dysgenic in modern societies, but Spencer was anti-imperialist whereas Royer "endorsed colonial expansion" (Hawkins 1998, 148). Nevertheless, there was a pattern among evolutionary thinkers to emphasize competition between groups in order to advance their tribal agendas. This pattern stands in sharp contrast to cosmopolitan visions like Darwin's expanding circles of sympathy in the *Descent*. Samuel Butler offers an original version of the pattern: his essay "Darwin Among the Machines," published the same year as *Man's Place in Nature*, argues passionately for exterminating machines before they evolve into a master race keeping the human species as pets (Butler 1914 [1863]). This external threat handily combines tribalism with panhuman sympathies.

Some evolutionary thinkers emphasized competition between individuals rather than groups and enlisted forces beyond nature to create harmony on a higher level of organization. Spencer fits partly into this category as well. His vision of human "evolution" was published before Darwin's *Origin*. It was not initially a biological theory, but an explicitly non-empirical attempt to discover the laws of the universe through logic (Spencer 1851). In *Social Statics*, Spencer starts from the premise that God has a benevolent intention for humanity and that we have internal forces developing us toward that intention. He predicts that "habit" will move us ever closer to the "social state," finally eliminating the struggle between individual and community. At the same time, he posits nearly absolute individual freedom as a prerequisite for happiness. Though he never abandoned his reasoning from religiously tinged "first principles," he used Darwin's natural selection to argue against societal interference—particularly against forcing education on children and giving succor to the "unfit" (Hawkins 1998, 82–103). Like Spencer, Huxley affirmed the natural inequality of individuals and the evolutionary advantage of intelligence (Huxley 2011 [1890]). In contrast to Spencer, however, he was firmly in favor of social institutions, charity, and public education (Huxley 2011 [1871]). He argued ambivalently that cultural forces could counteract the struggles inherent in "natural man" (Huxley 2009 [1893]). This ambivalence between natural inequality and salvation through education was inherited by his grandson. Julian Huxley combined a belief in natural inequality and the importance of eugenics with formulating UNESCO's first manifesto on global education as a means toward global peace (Huxley 1931, 1946).

Ernst Haeckel, arguably the most internationally visible proponent of Darwin's theory, held a position fairly close to that of the two Huxleys. He believed in natural inequality and competitive instincts (Haeckel 1879, 92–94). His social vision stood against the orthodoxy of his society and generally emphasized individual freedom: he was committed to scientific education, equal opportunities, tolerance of homosexuals, and a long-term vision of global peace; he was openly atheistic, equally critical of priests and backward scientific peers, and vocally against anti-Semitism (Richards 2009, 269–76, 327, 371–72, 489–512). He also opposed socialism. When his former teacher Rudolf Virchow attacked Darwin's theory by appealing to German anti-socialist sentiments, Haeckel responded that, if anything, Darwin's theory made socialism impossible:

> For the theory of descent proclaims more clearly than any other scientific theory, that that equality of individuals which socialism strives after is an impossibility, that it stands, in fact, in irreconcilable contradiction to the inevitable inequality of individuals which actually and everywhere subsists. (Haeckel 1879, 91)

For Haeckel, the individual variation of organisms on which natural selection depends precluded any full subordination of them to a community. Human happiness required individual freedom, including the freedom to pursue more exalted positions in society and receive proportional rewards: "in the constitutionally organised communities of men, as of the lower animals, neither rights nor duties, neither possessions nor enjoyments have ever been equal for all the members alike nor ever can be" (1879, 91). However, Haeckel did not glorify the competitive natural order, but rather saw it as a "cruel and merciless struggle" to which humanity was subjected: "We may profoundly lament this tragical state of things, but we can neither controvert it nor alter it" (93). Like Huxley, Haeckel accepted the struggle and suffering inherent in nature but found hope for harmony at a vaguely supernatural higher level. While professing metaphysical monism and scoffing at immortal souls, he injected his purely material universe with traditional German Romanticism: Nature, tragedies and all, was essentially a life-giving unity synonymous with God (Haeckel 1894, 1905; Richards 2009, 19–36).

2.4 Nature as Antagonist or Protagonist

As Haeckel's case indicates, we often involve nature as a moral agent in our social conflicts. Many conscientious attempts have been made to separate the realm of natural science from the realm of morality, including attempts by Haeckel himself, and by Huxley (Huxley 2009 [1893]; Richards 2009, 322–27). I believe there are two obstacles to any complete separation of that kind: (1) the fact that knowledge of what exists inevitably affects views of what can or should exist, and (2) the seemingly universal human tendency to personify natural forces as antagonists or protagonists. The first obstacle is logical: all suggestions about how human beings should act presuppose ideas of what human beings are capable of—in other words, what our nature is and what our place in nature is. Hence Haeckel's claim that we are incapable of thriving in socialist utopias because of natural inequality; hence also Wallas's argument that our dispositions change so slowly that society needs to be fitted to them, and Darwin's and Spencer's cases for infinite progress based on the idea that moral habits can be biologically inherited. The second obstacle is more insidious but perhaps less inevitable: the human mind working against itself.

The human tendency to enlist natural selection on our side or declare war against it is most likely an over-generalization of our tendency to shape social life into simple conflict narratives. As Darwin suggested, such "tribal" narratives sometimes have devastating effects, and often get in the way of objective knowledge, but they are also a core feature of our sociality: they provide psychological stability for individuals and unity for groups. The imagined antagonism between natural selection and society may gain further strength from other cognitive biases: more general dispositions to structure reality into categories for easy navigation, to distinguish ourselves from the rest of the world in order to form symbolic self-images, or to imagine intentions behind natural forces because it has been better to err on the side of caution in a world that is full of actual hostile intentions (Brown 2004, 2010; Geary 2007; Norenzayan et al. 2006; Turchin 2006). In any case, the moral valuation of natural selection is a mainstay of Darwinian thinking. Hawkins found it so pervasive in nineteenth- and twentieth-century social Darwinist literature that it provided the subtitle for his monograph: *Nature as Model and Nature as Threat*. His awareness of the tendency, of course, was not new. In the *Origin*, Darwin warns that "we may easily err on both sides" in perceiving good or evil in nature: we may go too far in "admiring with enthusiasm" its "inimitable contrivances" or in lingering with abhorrence on its "less perfect" aspect (2003 [1859], 221). Many early evolutionary thinkers err on one or the other side depending on the topic at hand, so that nature "vacillates between something to be emulated and something to be conquered" (Hawkins 1998, 269). Darwin's insight did not make him fully immune to this problem either.

For Darwin, as for many other Victorians, the antagonistic and protagonistic aspects of nature were caught up in Christian mythology. Part of the evidence leveled in the *Origin* against the Christian creation myth is the lack of a benevolent intention in nature. If God created each species to be perfectly fitted to its environment, it is difficult to account for the suicidal sting of the bee, the extreme waste of eggs by the Molothrus birds, or the imperfect correction for the aberration of light in "that most perfect organ, the human eye" (2003 [1859], 221). Natural selection, in contrast, is an imperfect natural mechanism fully consistent with these flaws. Though Darwin's case is logically sound, he invests the contrast between perfection and imperfection with a moral dimension. In his rhetoric, what is perfect is not just what is functional for the survival of an organism or a species, but what is beautiful and right according to our human sensibilities. He makes the opposite of divine perfection chillingly present:

> We need not marvel at the sting of the bee causing the bee's own death; at drones being produced in such great numbers for one single act, and being then slaughtered by their sterile sisters; at the astonishing waste of pollen by our fir-trees; at the instinctive hatred of the queen-bee for her own fertile daughters; at ichneumonidae feeding within the live bodies of caterpillars; and at other such cases. The wonder indeed is, on the theory of natural selection, that more cases of the want of absolute perfection have not been observed. (2003 [1859], 387)

In this light, natural selection appears to be not just an amoral natural force like gravity, but a force cruelly sporting with lives. The mechanism producing "death," "slaughter," "astonishing waste," "hatred," and "feeding within live bodies" looks like the antithesis of the benevolent force imagined by creationists. As Darwin said, natural selection made the instances of cruelty explicable, expected—even unexpectedly few. In wresting species creation from God, he thus had to dance and stretch to keep it from the hands of the arch-antagonist. He famously expressed that idea to his friend Hooker three years before the *Origin* was published: "What a book a Devil's chaplain might write on the clumsy, wasteful, blundering low & horridly cruel works of nature!" (1856). Though Darwin consciously avoided writing that chaplain's book, he had read it in his field work and experiments, and he could not avoid letting some of its imagery seep into the *Origin*.

Darwin's horror at nature's cruelty was companion to his vision of future harmony in the *Descent*. He followed a long tradition by seeing human moral reasoning and social development as salvation from nature's cruelty. That tradition is perhaps most famously embodied by Thomas Hobbes, but Huxley, aligning himself with the tradition, traced it back to Christian asceticism and the ancient Stoics (2009 [1893]). For Darwin, as we have seen, the expansion of sympathy was never a clear-cut case of society conquering nature. Sympathy had its roots in *natural* social instincts that had kept families and tribes together for survival. He believed that the moral habits were themselves naturally inherited. Nevertheless, he clearly believed that there would be progressive improvement as the natural state became civilized, and that cultural norms played an indispensable part in refining morality.

Whether one imagines nature as primarily antagonistic or protagonistic matters hugely for how one imagines the past and future of the human species. Darwin's vision was shared by many evolutionary thinkers who emphasized cooperation: those who were aware of the violence in nature, including human nature, but saw it as a gradually vanishing remnant of primitive stages of development. Such thinkers tended to stress the need for government guidance—education and long-term planning—if the human species were to overcome its violent past. Evolutionary thinkers who had an antagonistic view of nature but emphasized competition also stressed government guidance, but saw the suppression of violence and cruelty as an eternal struggle rather than a war that could be won. Such thinkers tended to view culture as the besieged protagonist. Huxley, for instance, oscillated between seeing nature as a gradual progression toward good and seeing culture as the tragic protagonist struggling against nature, with his complete separation of evolution and ethics as the most extreme expression of the tragic view (2009 [1893]). Huxley's student Wells, as we shall see in a later chapter, oscillated similarly between tragedy and heroic progression, but ended up developing an extreme version of the heroic progression with his utopian World State.

Those who saw nature as a protagonist and emphasized cooperation tended to claim that violence and cruelty had been introduced by social institutions. Thus, in their view, removing the artificial restraints of society would lead to natural peace and equality. Their visions for the future could be some variant of anarcho-communist utopias, or of noble-savage paradises in the vein of Jean-Jacques Rousseau (Hawkins 1998, 151–83). Those who saw nature as a protagonist but emphasized competition also saw social restrictions as a hindrance, but a hindrance to the progress of the elite rather than to equality and peace. Spencer's laissez-faire vision and the individualist philosophy of Friedrich Nietzsche are probably the most prominent examples of this position. However, there were more extreme versions at the margins of serious evolutionary thinking, like the 1896 manifesto *Might is Right, or The Survival of the Fittest*, published by the pseudonym Ragnar Redbeard and later echoed phraseologically—whether through direct influence or convergence—by Nazi ideologues (Redbeard et al. 2019 [1896]; Hawkins 1998, 278–84). Thus, social visions were fundamentally shaped by whether nature was seen as a victorious or an ever-besieged ally, as an invincible or a retreating opponent.

Perhaps more than any other evolutionary thinker, Darwin attempted to subsume moral conflicts within natural selection instead of the other way around. He struggled openly with the temptations to view natural selection as part of a moral system: to imagine it as a benevolent or malevolent force, to imagine it leading us away from or toward a natural state of battle-hardened power or social harmony. Though he may be said to lean more toward certain corners of these spectra, his thinking explored all corners. In the *Origin* and the *Descent*, he hints at natural selection's perceived satanic aspect and divine power to improve, responds with horror to its cruelty and waste, and marvels at its ingenuity and beauty. He believes that natural selection has produced vicious self-interest and the power to pursue it, but also that it has produced sympathy and reason and can combine the two toward moral progress. It is the mechanism shaping all benevolences and cruelties of the living world. Though he was ready to supplement it with Lamarckian inheritance of habits and refine it in accordance with new discoveries, he was also ready to extend its basic logic to every trait and behavior of every organism—from the wax cells of bees to the prejudices, arts, wars, and religions of human beings. The imaginative strain of such a unified view is summed up in a line of the *Origin's* oft-quoted final paragraph: "Thus, from the war of nature, from famine and death, the most exalted object which we are capable of conceiving, namely, the production of the higher animals, directly follows" (2003 [1859], 397–98). The principle of death implicit in natural selection is not just an ominous presence at the core of life, but its direct cause. Everything we consider exalted—everything beautiful and good and significant—derives from a process that cannot but be gruesome to our human sensibilities.

2.5 Making Meaning

The most general imaginative problem with natural selection is its failure to create a sense of purpose. All its particular imaginative difficulties come together into this problem: its abstract nature and astronomical scope, its implication of constant competition, and its decentralization of the human species (Bowler 1989). One might deal with this problem by refashioning natural selection into a cog in a political or moral system, or by rhapsodizing over the natural grandeur of the human species—or by attempting to banish evolution from the realm of purpose altogether. In the *Descent,* Darwin discusses the tendency of people to denounce natural selection as "irreligious" (1981 [1871], 396). He explains this tendency as a "revolt" of the human mind: "our minds refuse to accept" species or individuals as the result of "blind chance" (396). This, he claims, comes of the gradualism as well as the randomness of natural selection: a creation story fuelled by innumerable small accidents would not be satisfying even if we could imagine every individual accident as "ordained for some special purpose" (396). We prefer such a "grand sequence of events" to be more distinct, spectacular, and unified by an apparent purpose (396). In his view, this revolt of the human mind is a shortcoming. There is nothing logically more irreligious about explaining the origin of a species through a natural mechanism than there is about explaining "the birth of the individual through the laws of ordinary reproduction" (ibid.).

Given the acrimonious public debate of the late nineteenth-century, Darwin had a stake in sidelining the religious objections to his theory. Stressing the shortcomings of the human mind was therefore rhetorically convenient. His claim that there is nothing irreligious about natural selection might also look a bit disingenuous next to his open use of Christian creationism as a foil in the *Origin.* The Bible provided one account of how species emerged and Darwin provided another. Though Darwin claimed he was no more irreligious than scientists who explain reproduction, there is no Christian account of reproduction with which science clashes so directly. Nevertheless, the mental shortcoming he describes is real. Though he offered no explanation for why species and individuals are so "grand" to us that our minds demand them to be clearly purposeful, he affirmed the reality of the fact. Possibly, his affirmation was based on the responses he had gotten to his theory during more than a decade, and on the imaginative difficulties he had himself experienced with it. The problem of deriving a sense of purpose from natural selection is closely linked to the problem of maintaining a symbolic human self-image.

The positive and negative values contained in a self-image need to have reference to some larger system of meaning. Traits and behaviors can only have value from particular perspectives that identify specific functions and goals. In the symbolic self-image of *Man's Place*, which gained such enthusiastic approval from Darwin, Huxley portrayed human traits and behaviors as inherently valenced. He positioned himself against pedantic deductive moralism and ostensibly advanced a pragmatic morality based on natural health and beauty. However, the vocabulary he used to describe positive and negative human values was telling. In his vision, humanity will naturally strive to avoid not only "degradation," but "sin." Next to other animals we tower like the Alps, "raised by inward forces to that place of proud and seemingly inaccessible glory," so that travelers "can hardly discern where the deep shadowed crags and roseate peaks end, and where the clouds of heaven begin." The "philanthropist" and the "saint" will work against the "foundation" of our nature, which contains "all the selfish passions and fierce appetites of the merest quadruped." Our greatest achievement is to become "transfigured" from this quadrupedal nature by reflecting "a ray from the infinite source of truth." For most Victorians, the larger system of meaning was provided by the Christian faith. Good was good because it was close to the deity and would bring ultimate reward, and evil was evil because it departed from the deity and would bring ultimate punishment. As Darwin implies in the *Descent*, religion provides satisfaction partly by building on our social instincts. But it also provides unique satisfaction by picturing an ultimate intention behind the cosmos, and a role for humanity in that intention. Huxley's human self-image emphasizes the naturalistic core of religion—the amplification of social instincts toward virtuous health—and sneaks in the ultimate intention by association.

Darwin famously went through a loss of religious faith, complicated by his devout wife and his aversion to public debate (Darwin 1969 [1887]). Shortly after his death, the anti-clerical firebrand Haeckel published a letter in which Darwin admits to doubting the existence of an afterlife and not believing "that there ever has been any revelation" (quoted in Richards 2009, 351). This did not endear Haeckel to English authorities—who considered the publication an "indiscretion"—but it earned him the gratitude of those who wanted support for their atheistic systems of meaning (Richards 2009, 352). Haeckel's merging of Darwinism and

German Romanticism, which he called Monism, had by this time gathered a large following. It had developed into a fully fledged naturalistic faith not too far from Huxley's vision of natural human grandeur. As Haeckel put it, Monism was a concept of "Divine Nature," opening "an inexhaustible fountain of aesthetic enjoyment" to satisfy simultaneously humanity's "searching understanding" and "yearning spirit" (Haeckel 1894, 85–86). The idea of nature as a system meaningful in itself had great attractions, but it inevitably led back to the imaginative problems it was trying to solve. Our systems of meaning require specifically human meaning, and nature is not reducible to a human moral system.

Many evolutionary thinkers objected to Haeckel's division of science and religion (Richards 2009, 343–90). Instead of shaping naturalism to satisfy their need for a mythology, the religious among them tried to overcome the mental "revolt" identified by Darwin: they tried to imagine the guiding hand of the Christian god behind natural selection. Philosopher and historian John Fiske faced Darwin's challenge head on, asserting that the minute changes of evolution could be considered "an immediate manifestation of the creative action of God" (Fiske 1879, 7). He struggled to maintain this theist vision, sometimes imagining that natural selection was just a phase of the divine intention (Hawkins 1998, 107–8). The distinguished American palaeontologist Edward Drinker Cope attempted to reconcile his Darwinism with the Bible (1871, 33–36)—then, sixteen years later, admitted that his attempt had been mostly unsuccessful (1887, vii). Church minister and abolitionist Charles Loring Brace "saw no contradiction between Darwinism and a belief in God as creator" (Hawkins 1998, 63). However, Brace regarded natural selection as a struggle of individuals against their environment, with no competition between the individuals themselves (Hawkins 1998, 64–65). This made it possible for him to imagine natural selection as equivalent to his god: visiting sickness and death on "vice" and "indulgence," rewarding Christian perseverance, and working "always toward temperance and virtue" (quoted in Hawkins 1998, 64). Despite such heavy revision of Darwin's theory, Brace admitted that it was "difficult to recognise the hand of God in the play of chance and natural selection" (Hawkins 1998, 65). Similar attempts to reconcile creationism and evolution continued to be made throughout the twentieth century (Bowler 2007). They are still being made on a fairly large scale, in movements like Intelligent Design and through the theistic evolution officially recognized by the Catholic church. As Bowler notes, Darwinism has also been the focus of modern liberal intellectuals who seek to reconcile their religious faith with their respect for science (2007, 1–29).

The climax of Huxley's human self-image in *Man's Place* contains a hint of one of the most widely used goals for new systems of meaning: the growth of culture. Huxley's religiously tinted vocabulary brings his vision closer to that of an Enlightenment scientist unveiling truth than to a general intellectual participating in cultural self-fulfilment. Nonetheless, the image of language allowing us to rise "far above" other animals by accumulating our "experience" suggests that the accumulation of culture is itself a goal. By implication, future generations will be able to rise even further above the rest of nature: be wiser and more powerful, live more conscious lives in more controlled and harmonious environments. This is not just a vision of scientific progress. It is in tune with Matthew Arnold's attempt to make culture take the place of religion—an attempt that had a paradigmatic influence on humanist scholars (Arnold 1882). Though Arnold was not an evolutionary thinker, he was closely connected with Huxley, and he was a major figure in the intellectual environment during the Victorian loss of faith.

The purpose Arnold set for humanity was to learn the best that has been thought and said in the world, so that our own age could make its best possible contribution. In other words, we were to build our platform of experience ever higher, and become the most we could be. Huxley and Arnold debated the value of learning Greek and Latin, but they agreed that cultivation included knowledge of both the sciences and the arts, and that such cultivation was the foundation of a meaningful life (Arnold 1974 [1882]; Huxley 1893 [1880]). Though Arnold's purpose did not presuppose a benevolent intention watching over us, it was possible to invest culture with meaning the way Haeckel invested nature with meaning: seeing it as a benevolent purpose in itself. In fact, culture could be more easily imagined as an inherently meaningful system, since, unlike nature, it was shaped to fit the human mind in the first place. If one could locate the purpose of human existence in culture, one could also downplay the constraints of biology and focus on the developmental potential of art, technology, and moral consciousness. Thus, cultural growth was, and still remains, a compelling human purpose for many. Shaw, who did not give much credence to Darwinian evolution, maintained a version of conscious cultural evolution. According to his "new vitalism" or "creative evolution," will and tools could get us anywhere: "If you can turn a pedestrian into a cyclist, and a cyclist into a pianist or a violinist, without the intervention of Circumstantial Selection, you can turn an amoeba into a man, or a man into a superman, without it" (quoted in Hawkins 1998, 167). Wells, building on Huxley's teaching, found his ultimate purpose in a spiritualized cultural force that he called the Mind of the Race: an ever-growing deposit of wisdom to which anyone could contribute (Wells 1915). Oscar Wilde married Arnold's idea of cultural growth to Walter Pater's hedonistic aestheticism, arguing that our purpose was individual self-fulfilment (Wilde 2003 [1891]-a). All individuals who realized their full potential would contribute to the refinement of the Soul of Man—and one of his examples of such realization was Darwin (Wilde 2003 [1891]-b). It is tempting to call it a Wildean pose to turn the discovery of our lack of purpose into part of our purpose. However, Wilde's use of Darwin was perfectly in line with the tradition of investing cultural growth with mythological significance. Those who felt their faith in a supernatural overlord crumble turned their worship to the human achievements that had previously been considered offerings to the overlord. It is hard to believe that Huxley did not think of Darwinian evolution when he described humanity "reflecting, here and there, a ray from the infinite source of truth."

As Darwin noted, we do not derive a sense of purpose from pure logic, but from "grand sequences of events," or grand rhetoric on man's place in nature. Creating meaning in a human mind requires imaginative constructs that appeal to our emotions and establish intelligible relationships between concepts. That statement ceases to be tautological if one recognizes—as Darwin did—that our minds are limited. Despite the pooling of mental resources that produces occasional rays of truth, we struggle as a species to see parts of the cosmos that we have not needed to see to survive. Though culture transforms more smoothly than nature into an inherently meaningful system, the transformation has obvious problems. Some of the problems are the same as those for nature. In order to derive a sense of meaning from all of human culture, we need to simplify it: cancel out the mass of uninteresting or unappetizing facts, give it a direction that inspires action, and focus on spectacular features—in short, turn cultural history into mythology. Arnold's solution was to focus on a cultural canon. But like Huxley, he did not see human history as a straight line of progress, and he could not believe consistently in paradise at the end of the line. The transience of countless particular human cultures, and the fact that our entire cultural history covers an infinitesimal portion of natural history, also makes culture a fragile locus of meaning compared to nature. Haeckel's determined nature worship was in part a defence against this fragility. In the pantheistic vision, his young wife was not lost to him: she had returned to the stream of all living things, and after the end of civilization, the tentacles of medusa jellyfish could still recreate the locks of her hair (Richards 2009, 104–111).

Finding life meaningful, at its core, means being able to imagine possible human existences and reasons to pursue them. Our mythologies provide us with such meaning. However, the further we move into cultural and scientific awareness, the greater the pressure on our mythologies—pressure from alternative human mythologies, and ultimately from the non-human perspectives of the natural world. No early evolutionary thinker channeled the pressure of Darwinian evolution in a way that could re-establish the certain meaning of unchallenged mythology. No such certain meaning is possible when the implications of Darwinian evolution are drawn out in full. Our species simply has no discernible purpose or single intended way of life. However, evolutionary history probably means that mythologies will stay with us—at least until the human mind becomes radically different from what it seems to have been for around 50,000 years. Human beings have an extraordinary, though not complete, behavioral flexibility. In order to make that flexibility functional, we need to be imaginatively oriented to the core features of the human life cycle, and to the possibilities and dangers they entail in our particular cultural settings. Thus, myth-making serves a function. As indicated by the work of Darwin, Huxley, Haeckel, and others, we end up in myths despite ourselves. Taking this tendency seriously as a human need can only contribute to our understanding of ourselves, which is a prerequisite for controlling our biases and learning about reality. And of course, the lack of certain purpose does not mean a lack of truth. Even the wildest imaginative constructs can illuminate aspects of reality and guide us into individual or collective existences that satisfy actual human needs. The arts generally create wilder imaginative constructs than political, philosophical, and scientific writing, but they are also generally better at providing vivid imaginative experience to infuse our mythologies with meaning. The first literary authors who engaged with Darwinian evolution made extraordinary use of this capacity.

References

Arnold, Matthew. 1882. *Culture and Anarchy: An Essay in Political and Social Criticism*. London: Smith, Elder & Co.

———. 1974 [1882]. "Literature and Science." In *The Complete Prose Works of Matthew Arnold*, edited by Robert H. Super, 425–30. Ann Arbor: University of Michigan Press.

Bebel, August. 1910. "Darwinism and the Condition of Society." In *Woman and Socialism*, 253–58. New York: Socialist Literature Co.

Boehm, Christopher. 1997. "Impact of the Human Egalitarian Syndrome on Darwinian Selection Mechanics." *The American Naturalist* 150 (S1): 100–21. https://doi.org/10.1086/286052.

Bowler, Peter J. 1989. *Evolution: The History of an Idea*. Berkeley; Los Angeles; London: University of California Press.

———. 2007. *Monkey Trials and Gorilla Sermons: Evolution and Christianity from Darwin to Intelligent Design*. Cambridge; London: Cambridge University Press.

Brooke, John Hedley. 2014. *Science and Religion: Some Historical Perspectives*. Cambridge: Cambridge University Press.

Brown, Donald E. 2004. "Human Universals, Human Nature & Human Culture." *Daedalus* 133 (4): 47–54. https://doi.org/10.1162/0011526042365645.

———. 2010. "The Universal People." In *Evolution, Literature, and Film: A Reader*, edited by Brian Boyd, Joseph Carroll, and Jonathan Gottschall, 83–96. New York: Columbia University Press.

Butler, Samuel. 1914 [1863]. "Darwin Among the Machines." In *A First Year in Canterbury Settlement with Other Early Essays*, edited by R. A. Streatfeild, 179–85. London: A. C. Fifield.

Carroll, Joseph, Jonathan Gottschall, John Johnson, and Daniel J. Kruger. 2012. *Graphing Jane Austen: The Evolutionary Basis of Literary Meaning*. Cognitive Studies in Literature and Performance. New York: Palgrave Macmillan.

Cope, Edward Drinker. 1871. *On the Hypothesis of Evolution: Physical and Metaphysical*. New Haven: Chales C. Chatfield & Co.

———. 1887. *Origin of the Fittest: Essays on Evolution*. New York: D. Appleton and Company.

Darwin, Charles. 1854. "Letter to T. H. Huxley on 2 September 1854." *Darwin Correspondence Database*. "Letter no 1587." University of Cambridge. http://www.darwinproject.ac.uk/entry-1587.

———. 1856. "Letter to J. D. Hooker on 13 July 1856." *Darwin Correspondence Database*. "Letter no. 1927." University of Cambridge. https://www.darwinproject.ac.uk/letter/DCP-LETT-1924.xml.

———. 2003 [1859]. *On the Origin of Species by Means of Natural Selection*. Edited by Joseph Carroll. Petersborough: Broadview Press.

———. 1863, "Letter to T. H. Huxley on 26 February 1863." *Darwin Correspondence Database.* "Letter no. 4013." University of Cambridge. https://www.darwinproject.ac.uk/letter/DCP-LETT-4013.xml.

———. 1981 [1871]. *The Descent of Man and Selection in Relation to Sex.* Princeton: Princeton University Press.

———. 1969 [1887]. *The Autobiography of Charles Darwin: 1809–1882; with Original Omissions Restored.* Edited by Nora Barkow. New York: Norton & Company.

Dewey, John. 1922. *Human Nature and Conduct: An Introduction to Social Psychology.* New York: Henry Holt and Company

———. 1958. *Art as Experience.* New York: Capricorn Books.

Fiske, John. 1879. *Darwinism and Other Essays.* London and New York: Macmillan and Co.

Geary, David C. 2007. "The Motivation to Control and the Evolution of General Intelligence." In *The Evolution of Mind: Fundamental Questions and Controversies*, edited by Steven W. Gangestad and Jeffry A. Simpson, 305–13. New York: Guilford Press.

Ghiselin, Michael T. 2003. *The Triumph of the Darwinian Method.* Mineola: Dover Publications.

Gottschall, Jonathan. 2012. *The Storytelling Animal: How Stories Make Us Human.* Boston: Houghton Mifflin Harcourt.

Griskevicius, Vladas, Stephanie M. Cantú, and Mark van Vugt. 2012. "The Evolutionary Bases for Sustainable Behavior: Implications for Marketing, Policy, and Social Entrepreneurship." *Journal of Public Policy & Marketing* 31 (1): 115–28. https://doi.org/10.1509/jppm.11.040.

Haeckel, Ernst. 1879. *Freedom in Science and Teaching.* New York: D. Appleton and Company.

———. 1894. *Monism as Connecting Religion and Science: The Confession of Faith of a Man of Science.* London: A. and C. Black.

———. 1905. "God and the World." In *The Riddle of the Universe at the Close of the Nineteenth Century*, 275–91. New York and London: Harper & Brothers Publishers.

Haselton, Martie G., and Daniel Nettle. 2006. "The Paranoid Optimist: An Integrative Evolutionary Model of Cognitive Biases." *Personality and Social Psychology Review* 10 (1): 47–66. https://doi.org/10.1207/s15327957pspr1001_3.

Hawkins, Mike. 1998. *Social Darwinism in European and American Thought, 1860 1945: Nature as Model and Nature as Threat.* Cambridge: Cambridge University Press.

Huxley, Julian. 1931. "The Vital Importance of Eugenics." *Harper Monthly Magazine*, August.

———. 1946. "UNESCO: Its Purpose and Its Philosophy." *Preparatory Commission of the United Nations Educational, Scientific and Cultural Organization*, June 15.

Huxley, T. H. 1854. "Vestiges of the Natural History of Creation. Tenth Edition." *British and Foreign Medico-Chirurgical Review* 13: 425–39.

———. 2009 [1863]. *Evidence as to Man's Place in Nature*. Cambridge: Cambridge University Press.

———. 2011 [1871]. "Administrative Nihilism." In *Collected Essays: Volume 1, Methods and Results*, 251–89. Cambridge: Cambridge University Press.

———. 1893 [1880]. "Science and Culture." In *Collected Essays: Volume 3, Science and Education*, 134–59. London: Macmillan.

———. 2011 [1890]. "On the Natural Inequality of Man." In *Collected Essays: Volume 1, Methods and Results*, 290–335. Cambridge: Cambridge University Press.

———. 2009 [1893]. *Evolution and Ethics: Delivered in the Sheldonian Theatre, May 18, 1893*. Cambridge: Cambridge University Press.

Mayr, Ernst. 1991. *One Long Argument: Charles Darwin and the Genesis of Modern Evolutionary Thought*. Cambridge, MA: Harvard University Press.

Norenzayan, Aran, Scott Atran, Jason Faulkner, and Mark Schaller. 2006. "Memory and Mystery: The Cultural Selection of Minimally Counterintuitive Narratives." *Cognitive Science* 30 (3): 531–53. https://doi.org/10.1207/s15516709cog0000_68.

Orwell, George. 1941. "Wells, Hitler and the World State." *Horizon* 4 (20): 133–38.

Redbeard, Ragnar, Trevor Blake, and Peter H. Gilmore. 2019 [1896]. *Might Is Right, or The Survival of the Fittest: The Authoritative Edition*. Baltimore: Underworld Amusements.

Richards, Robert J. 2009. *The Tragic Sense of Life: Ernst Haeckel and the Struggle Over Evolutionary Thought*. Chicago: University of Chicago Press.

Spencer, Herbert. 1851. *Social Statics*. London: Chapman.

Stillweli, Arlene M., and Roy F. Baumeister. 1997. "The Construction of Victim and Perpetrator Memories: Accuracy and Distortion in Role-Based Accounts." *Personality and Social Psychology Bulletin* 23 (11): 1157–72. https://doi.org/10.1177/01461672972311004.

Turchin, Peter. 2006. *War and Peace and War: The Life Cycles of Imperial Nations*. New York: Pi Press.

Wells, H. G. 1975 [1895]. "Bio-Optimism." In *H. G. Wells: Early Writings in Science and Science Fiction*, edited by Robert M. Philmus and David Y. Hughes, 206–10. Berkeley: University of California Press.

———. 1975 [1897]. "Morals and Civilisation." In *H. G. Wells: Early Writings in Science and Science Fiction*, edited by Robert M. Philmus and David Y. Hughes, 220–28. Berkeley: University of California Press.

———. 1915. *Boon, the Mind of the Race, the Wild Asses of the Devil, and the Last Trump; Being a First Selection from the Literary Remains of George Boon, Appropriate to the Times, Prepared for Publication by Reginald Bliss. With an Ambiguous Introduction by H. G. Wells*. London: T. F. Unwin.

Wilde, Oscar. 2003 [1891]-a. "The Critic as Artist." In *The Complete Works of Oscar Wilde*, edited by Merlin Holland, 1108–55. London: HarperCollins Publishers.

———. 2003 [1891]-b. "The Soul of Man Under Socialism." In *The Complete Works of Oscar Wilde*, edited by Merlin Holland, 1174–97. London: HarperCollins Publishers.

Willard, Aiyana K., and Ara Norenzayan. 2013. "Cognitive Biases Explain Religious Belief, Paranormal Belief, and Belief in Life's Purpose." *Cognition* 129 (2): 379–91. https://doi.org/10.1016/j.cognition.2013.07.016.

Young, David. 2007. *The Discovery of Evolution*. Cambridge: Cambridge University Press.

CHAPTER 3

Darwinism in Literature

> The remotest discoveries of the Chemist, the Botanist, or Mineralogist, will be as proper objects of the Poet's art as any upon which it can be employed, if the time should ever come when these things shall be familiar to us … manifestly and palpably material to us as enjoying and suffering beings.
>
> —William Wordsworth, Preface to *Lyrical Ballads*

> The men of genius propound solutions which grow into systems of Theology or of Philosophy, or veiled in musical language which suggests more than it asserts, take the shape of the Poetry of an epoch.
>
> —T. H. Huxley, *Man's Place in Nature*

My argument about the clash between evolution and the human imagination assigns a special place to the arts. Though Darwin and Huxley occasionally produced rhetoric that rang in people's imaginations, they were not writing novels or poetry. They were advancing arguments—citing facts, following out logical sequences, and attempting to bring their readers to quite specific conclusions. Though they made some use of satire and personification, they did not create fictional characters who interacted and simulated complex social experiences for their readers. They sprinkled powerful lyrical images through their texts, but they did not compose extended musically modulated patterns of such images. The *Origin* and

E. Jonsson, *The Early Evolutionary Imagination*, Cognitive Studies in Literature and Performance, https://doi.org/10.1007/978-3-030-82738-0_3

Man's Place make use of hypothetical scenarios but do not take place in fictional worlds. They contain illustrative anecdotes, speculative narration, and hints of agonistic structures, but they could not be entered into narratological charts identifying three-act structures, hero's journeys, catharses, or relationships between fabula and syuzhet. In short, they do not create the sort of imaginative experience that enables the arts to guide human behavior in a particularly powerful way. This distinction between argument and verbal art is no more absolute than the Darwinian distinction between species. Nor is it any less of a fact. Species may diverge and converge over time, share traits and lack any single feature to mark them as unique. But at any given time, a species is a statistical reality, sharing a unique combination and gradation of traits, filling a specific niche. My study centers on the niche of verbal art.

The human imagination faced Darwin's theory in scientific arguments and political essays as well as in novels and poems. Though the non-fiction writing was generally less "veiled in musical language" and more constrained by the public responsibility of truth claims, non-fiction writers did as we have seen produce sparks of mythology. Literary authors not only engaged with the theoretical ideas of Darwin, Huxley, Spencer, and Haeckel, they also mirrored and expanded their evolutionary myth-making. They extrapolated the ideas into fictional worlds and kindled the imaginative sparks into blazes. If Darwin hinted at the possibility of progressive moral evolution and reflected ambivalently on the artificial selection of civilization, literary authors created utopian and dystopian societies populated by excruciatingly half-human shapes. If Huxley swelled rhetorically with the greatness of humanity and the fragility of civilization, literary authors led readers to heroic conquests of nature or through harrowing experiences that made civilization quake under their feet. If Spencer clinically advised present ruthlessness in favor of future harmony, literary authors gave Ruthlessness a face and name and made readers writhe bleeding in its grasp. If Haeckel speculated about pre-linguistic ancestors, literary authors gave them lives and shaped their experiences into tragedies and comedies; if he translated his grief into a vision of a spiritually unified nature, literary authors evoked brimful sunny meadows contrasted with the suffering individuals walking across them. Many such early stories and images have entered our cultural heritage, forming part of our new naturalistic mythologies. All of them show minds like ours facing an imaginative problem that has been passed down to us unsolved.

Like non-fiction writers, literary authors were preoccupied with the human self-image, saw nature as either essentially harmonious or essentially combative, envisioned agonistic relationships between nature and culture, and created structures of cosmic meaning. Some saw the human species as essentially peaceful and found cosmic meaning in working against corrupting culture to become one with harmonious nature. Some found cosmic meaning in resisting blood-soaked nature by supporting a fragile culture kept alive by those who represent the human spirit at its best. These preoccupations, differently combined, represent elemental problems for the human imagination when dealing with Darwinian evolution. Across cultures, we are intensely concerned with personal identity and group identity. Our minds register dichotomies between the human and non-human: psychology and biology, culture and nature, society and wilderness, the controlled and uncontrolled. Our emotions about such dichotomies are moulded by the most ancient factors of human social life beyond the family: hierarchies of domination and egalitarian coalitions. Because we have simultaneous impulses toward establishing personal power and toward being part of egalitarian coalitions, we want power for our own kind while portraying our own kind as general benefactors (Carroll et al., 2012). And we extend such visions of ourselves into the past and future, creating narratives of ideal development or heroic struggle that give us a sense of meaning. We have amassed these emotional and conceptual dispositions around the human life cycle over millions of years. They move us to define ourselves, single out our cultural practices from the rest of nature, see the world as all cooperation or all competition, and create cosmic narratives with ourselves as protagonists.

These cross-cultural imaginative tendencies establish a basic structure for understanding the literary myths of evolution. The next level in the structure is that of genres—the niches within the niche of verbal art. Thinkers from the time of Aristotle have claimed that genres are either universal or culturally specific. I suspect that genres will turn out to range from the universal to the culturally specific, with genres centering on basic emotions at the universal end and highly formalized treatments of culturally specific subjects at the other end. Since genre is a major parameter for the communication between authors and audiences, a fully developed and empirically tested genre theory would be invaluable for understanding how verbal art is created and experienced. However, until there is such a theory, the best option is reasoned analytic description of genres informed by both humanist genre theory and the evolutionary social sciences.

Since I focus on novels, I already work within a genre with a culturally specific origin (Watt 1957). The novel genre contains extended written works in free form that often have complex storylines and give access to the internal states of characters; it was pioneered by the episodic work of Miguel de Cervantes Saavedra and the epistolary tales of Samuel Richardson, and developed rapidly across cultures during the last three centuries. The novels I discuss appear around the middle of those three centuries in geographical hotspots for the development (Britain and the USA). Within that historical period, I zero in on novels that engage with a specific subject—Darwinism. Finally, among the novels that engage with Darwinism, I describe a spectrum of genres based on function and complexity. At one end of the spectrum are novels meant primarily as short-term entertainment, using simple thematic and stylistic structures to appeal to simple emotions: successful adventure stories with more or less lasting power. At the other end of the spectrum are novels exploring moral and philosophical questions in a nuanced way, using complex thematic and stylistic structures to develop complex emotions: tortured existential meditations that occupy the highest reaches of the cultural canon. In between the two poles are novels of ideas and fairy tales with didactic purposes, as well as adventure stories with elements of political allegory or hints of moral complexity. My analysis will traverse this spectrum from the simplest to the most complex.

The novels I discuss are not a scientific sample, but nevertheless a selection meant to be representative. Since my subject is the novelistic response to the clash between the human imagination and Darwin's theory, I exclude the bulk of early satire that set up evolution in the vaguest terms to ridicule it or accuse it of malice—doctrinaire dismissals like Wilkie Collins's *The Law and the Lady* and Catherine Liddell's *Margaret* (Henkin 1963 [1940], 76–94). There can be no clash with the imagination if the theory is reduced to a gorilla with a potty mouth about religion. I focus on the novels that portrayed Darwin's ideas, extrapolated from them, or engaged with them in ways that show some understanding of their import—however much they were subsequently embellished or cut into shape to fit the human mind. To see the most powerful imaginative tendencies, I deal with novels that either received special attention in their time or have received special attention since.

Literary meaning is not created in an abstract way by biology or culture, but by individual authors in communication with readers. This process is both extraordinarily complex and unpredictable. Literary works rarely evoke full and coherent worldviews. As Huxley put it, verbal artists "suggest more than they assert" (2009 [1863], 72). Or, as William Gladstone held when critiquing Mrs. Humphry Ward's epic of religious crisis *Robert Elsmere*, a novel can "put upon us what is not indeed a treatise, but more formidable than if it were":

> For a treatise must nowhere beg the question it seeks to decide, but must carry its reader onward by reasoning patiently from step to step. But the writer of the romance, under the convenient necessity which his form imposes, skips in thought, over undefined distances, from stage to stage, as a bee from flower to flower. A creed may (as here) be accepted in a sentence, and then abandoned in a page. (quoted in Ward 2013 [1888], 620)

The imaginative experience of a novel is not reducible to the scientific, philosophical, or political system behind it, precisely because the novel provides an experience different from reading a treatise on the system. There often may not be a consciously articulated or consistent system in the author's mind. Readers may not be aware that they are engaging with self-images, evaluating political causes or modifying their worldviews. But to quote Mrs. Ward herself, a reader may still feel a novel "in his veins like wine" (Ward 2013 [1888], 612–13). Novels may, without a reader's conscious awareness, expand the walls of a world that has seemed "sickeningly narrow and empty" to let in "crowd, excitement, and delight," introduce him to fictional characters who make him "take sides, compare himself with them, join in their fights and hatreds, pity and exult with them"—novels may even "set his thoughts, endowed with a new heat and nimbleness, to the practical unravelling of a practical case" (613).

However intricate, unconscious, and irrational the experience of literature, it is possible to rationally analyze the imaginative world a novel creates for its author and readers. What crowds move through it? What sides are depicted, what are the fights and hatreds, pities and exultations, desirable or disastrous practical ends? The answers to these questions necessarily include the style and structure of a novel, because its narrative form, tone, vocabulary, and network of references are inherent to the imaginative experience it provides. All these variables—including the nature of the crowds, hatreds, pities, and ends—are modulated by cultural setting and authorial temperament. The way readers are affected in turn depends on their cultural context, personalities, and mental capacities.

Thus, the basic parameters of our evolved psychology give rise to astonishing complexity. But our evolved psychology makes up the timber of the minds in which all literary meaning takes place.

3.1 Previous Literary Scholarship

My evolutionary perspective puts me at odds with most previous scholarship on Darwinism in literature. The conflict is sometimes partial, resulting from differences in purpose rather than in principles and methodology. Beginning in the early twentieth century, there has been a tradition in literary study of collecting Darwinian influences somewhat the way naturalists in the early nineteenth century collected beetles. Scholars in this tradition have not attempted much explanatory synthesis, but they have described, sorted, and classified specimens in ways that help create a full view of the subject. The greatest difference between them and myself is that I aim to explain the cross-cultural fascination of Darwinism and the particular ways it entered literature. Some scholars of this tradition might object to my use of an evolutionary framework for explanation—or to the aim of explanation itself—while others might not. Regardless, their transparent methodologies and evidence-based arguments make it possible for me to build on their work.

There is a more serious conflict between my work and the later tradition of literary scholarship on the Darwinian influence. Since the theory revolution of the later decades of the twentieth century, literary scholars have tended to use theories of discourse to explain both evolution and the depiction of evolution in novels. Rather than using evolutionary science to explain the human behavior of myth-making, they treat evolution as myth and pose their discourse theories as explanations for human behavior. On that basis, they make claims about the cause and effect of human self-definition that often conflict directly with an evolutionary explanatory framework. Because their research is guided by what I consider erroneous explanatory claims, and because their style favors performance over transparency, these scholars are most useful to me as polemical foils. Nonetheless, their work is like mine in recognizing that Darwinism is a special challenge to humanity and in seeking explanations for that challenge. They also produce evolutionary myths of their own, and thus form part of the continual imaginative preoccupation with evolution that is my subject.

One of the earliest and best examples of a reference collection, first published in 1940, is Leo Henkin's *Darwinism in the English Novel*. Henkin displays a daunting range of literary specimens: from Erasmus Darwin's zoological poetry and the jabs at Chambers's pre-Darwinian evolution in Benjamin Disreaeli's *Tancred*, through the initial vitriolic satire by religious critics like Marie Corelli, to the hobbled attempts at spiritual reconciliation like Charles Kingsley's *Water Babies*, the romances about prehistory like Henry Curwen's *Zit and Xoe*, and the detached end-of-century visions of human decline by H. G. Wells and Thomas Hardy. Henkin vividly evokes the historical context and includes a full description of natural selection (1963 [1940], 45–61). Outside the presentations of these specimens, he establishes no consistent analytic perspective. In the classic manner of early twentieth-century humanists, he evokes as much as he analyzes: he enters into each author so fully that he sometimes seems to become the person he describes. One moment he seems to dismiss Lamarckism (59–61), the next he seems to credit it (210–12); one moment he produces high rhetoric in praise of Huxley's battle against "obscurantist tactics" (68–75), the next he concurs that the obscurantist Samuel Butler was Darwin's "most brilliant" critic (209). Because he wrote shortly after the modern synthesis, he also makes claims that would later turn out to be gross scientific inaccuracies: for instance, he joins geneticists Bateson and De Vries in minimizing the role of natural selection, defines evolution as "progressive development to higher and more complex forms" (61), and puts the Piltdown hoax next to legitimate paleoanthropological finds (168–69). The solidity of his collection holds up under all these defects, largely because his own commentary on it is minimal.

Henkin's combination of solid scholarship and chameleonic analytic perspective is representative. The same old school humanist model has been applied to the Darwinian imagination by scholars through the twentieth century and up to the present. What is exceptional about Henkin—apart from his pioneering role, his thoroughness, and his occasional panache—is his modesty. In his introduction, he makes his one attempt to explain the novelistic clash between Darwinism and Victorian faith, identifying three causes: "the interest of Victorians in religion," "the Victorian passion for science" and the fact that "the novel continued to be the chief medium of cultural communication" (9–10). As he sees it, the explanation for one local cultural phenomenon is simply the combination of three other local cultural phenomena. Causes and effect are accidents of history mostly incomprehensible outside England's vicinity between

1860 and 1910. He concludes that the greater part of his subject is "the shadow of a shadow," "cold as a fossil," and that most of the specimens he is about to display are "of historical interest merely" (10, 11). This assessment, which in my view minimizes the subject to the point of surrealism, forms an amusing contrast to the more common tendency to make unearned claims of universal significance for one's subject. In fact, a claim of universality had already been made for Henkin's subject by its other major pioneer, Lionel Stevenson.

Compared to Henkin's modest tour through a brimful naturalist's collection, Stevenson's 1932 *Darwin Among the Poets* is a majestic sermon on the morphology of a single genus. Its characteristic commanding tone is struck in the introduction:

> The irruption of science is probably the most interesting phenomenon in nineteenth-century literature; and indeed that is only one chapter of the change which was then affecting every province of human thought, and of which the reverberations still surround us. (1963 [1932], 1)

The "change" of which he speaks is the assimilation of a scientific worldview. Stevenson registers the immense consequences of that assimilation, and spots reverberations in everything from a 1926 novel by H. G. Wells to contemporary poets and "the grievance of the New Humanists against modern literature" (1, 1–6, 298–345). The Victorian poets' mission to "reconcile the new principle with some sort of spiritual system" forms a "gulf" between them and "all antecedent poets"—a gulf that is "probably the most abrupt in the whole record of human thought" (6). These are fighting words worth the tussle. But Stevenson does not argue his case or attempt to account for the extraordinary gulf. Instead, he settles into tracing particular imaginative ideas about science that were supposedly originated by Victorian poets, and that he claims survive uncredited in twentieth-century culture (2–54). As a result, he appears to be arguing that the poetic ideas themselves caused the gulf, the change, and the reverberations. Like Henkin's "interest in science and religion," they are local cultural phenomena producing other cultural phenomena—only a little less locally.

There is a clear contrast between Henkin's self-effacing minimizing of his subject and Stevenson's majestic claim for its universality; a contrast that must have been glaring when both monographs appeared in reprint in 1963 (ten years after the Piltdown hoax was revealed without a ripple of revision in Henkin's text). Nevertheless, the two pioneers have more in common than separates them. In Stevenson's conception, the ideas he traces do not appear to be invented, passed on to, or independently arrived at by human minds in contact with a larger world. The ideas are born and develop, have conflicts, die or couple and form family histories—like life forms whose evolution we can study though their First Cause remains outside our ken. Thus, his brief evocation of an immense scope does not prevent him from being, in essence, a classifier and a collector. He stops short at a cultural history that seems to exist independently of human psychology and the material world, leaving his hints at universal significance as local rhetoric.

Despite the gulfs, changes, and reverberations in literary study as a discipline, the model of Henkin and Stevenson persists up through the twenty-first century. Mark Hillegas's *The Future as Nightmare* (1967) traces the influence of H. G. Wells's evolutionary ideas on the dystopian novelists of the early twentieth century, while Steven McLean's *The Early Fiction of H. G. Wells* (2009) traces the influence of scientific ideas on H. G. Wells. Tom Gibbons's *Rooms in the Darwin Hotel* (1973) is "mainly concerned with certain relationships between ideas and literary criticism during the period 1880–1920" (viii). Redmond O'Hanlon's 1984 *Joseph Conrad and Charles Darwin: The Influence of Scientific Thought on Conrad's Fiction* identifies its scope in its title, as does Nicholas Ruddick's *The Fire in the Stone: Prehistoric Fiction from Charles Darwin to Jean M. Auel* (2009). In 2012, Michael R. Page tracked the family histories of Romanticism and Darwinism through the nineteenth century, calling them "the two overarching myths for the modern scientific imagination," "bound together in our cultural consciousness" (4). I am indebted to most of these works for their solid scholarship, but none of them has produced an explanatory framework beyond the influence of culture on culture.

Despite publishing in the latter half of the same century, the currently most prominent authorities on evolution in Victorian literature are virtually disconnected from the tradition of Henkin and Stevenson. Gillian Beer, whose 1983 *Darwin's Plots* remains the core reference, mentions none of the early literary scholarship. Her respectful second George Levine was more circumspect when publishing *Darwin and the Novelists* in 1988. While paying generous tribute to Beer, he acknowledges in an endnote that Henkin and Stevenson wrote "perhaps the best known full-scale general studies" of the topic; but he dismisses Henkin's study as "an extended catalogue of references" and Stevenson's as "very different in objective and argument from mine" (1991 [1988], 275). He shows similar circumspection when prefacing Beer's *Darwin's Plots*, by at once connecting her to the previous scholarship and justifying her disconnect from it. According to Levine, Stevenson made a "useful," "pioneering," and "important" contribution to Beer's general topic, but Stevenson's work can be set aside because it focuses "almost entirely on Darwin's 'influence' on later writers" (Beer 2009 [1983], xii–xiii).

What puts Beer's contribution in a category of its own, Levine explains, is that she attends "to the texture of Darwin's language and its deep historical roots"; she recognizes "how Darwin's work transformed the fundamental myths of the culture, myths upon which its language was built and whose vestiges help give Darwin's writing its capacity to escape monolithic meaning" (xiii). Apart from the cryptic end to this sentence, it might be hard to tell from it how Beer's study differs from previous scholarship—say, from Stevenson's argument about our cultural debt to Darwin's imaginative impact, or from Henkin's rich historical contextualization of the Darwinian debates. But Levine is not wrong to identify a difference. As the end of his sentence hints, Beer's purpose is to make Darwin "escape monolithic meaning": to show that Darwinian evolution both means an endless number of different things and reveals the truth that meaning is indeterminate. In order to support these contentions, she engages in intricate verbal play. As Levine puts it, her language "makes it impossible for readers to relax, for it forces them to see that language does not hold still, neither Darwin's nor her own"—her words "serve not only the obvious utilitarian effect of getting it right but of getting it right in ways that expand possibilities and intimate abundance" (xiii). In other words, Beer's purpose stands above and apart from the "obvious utilitarian" purposes of Henkin and Stevenson. Levine considers it a chief merit of her book that the ambiguity of her language mirrors her argument for ambiguity, and that she expresses the truth of her subject in a way that stresses multifariousness. This distinction between her work and that of Henkin and Stevenson does not in my view excuse her from mentioning them, but it explains why she does not. Her exclusion of the early scholarship is part of a general tendency to privilege mind-expanding performance toward a foregone conclusion over transparent scholarship and evidence-based argumentation.

Both Beer and Levine proceed from standard poststructuralist assumptions that our minds cannot reach outside culture, so that our theories about ourselves and the world are arbitrary results of our cultural histories. Rather than proclaim this view outright, they use a common form of rhetoric that consists in switching between an exciting but unsustainable argument and an undeniable but unexciting argument (Carroll 1995, 56–68, 69–70). From the beginning of her book, Beer suggests that Darwin's theory is more fiction than truth: it is "first a form of imaginative history," and it "cannot be experimentally demonstrated sufficiently in any present moment" (Beer 2009 [1983], 6). When formulating his theory, Darwin rearranged "the elements of creation myths, for example substituting the ocean for the garden" (7). The goal of his Frankensteinian rearrangement was "telling a new story": a story that "proved not to be simple or single" (3).

The unavoidable implication of these claims is that Darwinian evolution is predominantly an inspired myth—a suggestive, ambiguous fantasy most notable for turning old tropes against their traditions. In a preface to the second edition of her book, Beer denies having implied any such thing. She did not mean "that Darwin's work is 'fiction,' as some puzzled readers at first assumed" (xxv). The ("monolithic"?) meaning of her book is only that Darwin's culture-laden language "is not a layer that can be skimmed off without loss" (xxv). Her aim was to analyze Darwin's use of metaphor and culturally prominent phrases, to demonstrate "the degree to which narrative and argument share methods" (xxv). This defence, consisting of undeniable statements, gets her dangerously close to the toothless version of her argument: Darwin's language was influenced by his culture. The only part of this statement that would have surprised Henkin is that it is framed as an argumentative claim. And so, Beer reintroduces the suggestion that evolution is fictional by ending on a rhetorical question: "indeed," her book "enquires what differences can be maintained between narrative and argument" (xxv). This makes an expert completion of the cycle: from hedged heavy implications that evolution is fiction, to a retreat when challenged into the obvious truth that Darwin was influenced by his culture, to the suggestion that scientific arguments are indistinguishable from fictional narratives.

Levine, politically describing the controversy without involving himself, mirrors Beer's movement in his preface to her book. He portrays her "bemusedly" explaining that she does not claim that Darwin's theory is fiction:

> But her point, richly made over and over again, is that the language in which Darwin's theory is articulated is thick with the culture in which Darwin lived and that fully to understand the 'science,' one must recognise how the language contributed to it, evoked resistances, entailed compliances. (Beer 2009 [1983], xii)

The richly made point is, of course, the innocuous idea that Darwin was influenced by his culture. But before the sentence has ended, Levine re-inflates that idea into its tigerish poststructuralist version: It is necessary to study Victorian culture in order to understand Darwin's theory *as science*. One cannot comprehend natural selection without being a Victorianist. This idea follows naturally from the poststructuralist assumption that all theories are arbitrary products of history, true only insofar as they imply the impossibility of knowledge. Levine's claim is big and exciting, with consequences not just for literary study but for numerous other disciplines and even for the world of action. But the most generous interpretation of *fully* and *science* could not make it true. If challenged, Levine would no doubt expand those two words to bring himself back to the obvious version of his argument. Of course he never suggested that biologists must become Victorianists. Are his critics suggesting that Darwin lived in Olympian isolation from Victorian culture?

Levine exercises the same rhetorical capacities in his own book. He simultaneously reserves judgment and takes a stance on whether science is fiction. That is, he does not want to plunge "into the wars among philosophers and historians of science" but he nonetheless ("for the purposes of this volume") "considers science as an unprivileged form of cultural discourse, 'a cultural formation like any other'" (1991 [1988], 5). He carefully avoids naming individuals he disagrees with, but he characterizes his opponents in a roundabout way by aligning himself with those who challenge "notions of science as progressive, empirical, verifiable and falsifiable, objective, and uncontaminated by the winds of interest or subjectivity" (5). This characterization of his opponents subtly blends an actual position with a false dichotomy. Anyone who credits science would admit that it is "progressive, empirical," and at least "falsifiable" at certain levels if not "verifiable" in any ultimate sense. No sensible person—today or in the nineteenth century—would claim that science is "uncontaminated by the winds of interest or subjectivity." The scientific method is a systematic attempt to lessen the force of such winds precisely because they have been recognized as a constant threat. As the previous chapter showed, Darwin was often concerned with his cognitive and cultural biases. But since Levine is arguing for the power of subjectivity (partial or complete depending on the version), he portrays his opponents as believers in complete objectivity.

If a reader catches the elusive words of Beer and Levine long enough to see the false suggestion that evolution is fiction, they can retract their implications and retreat into the undeniable truth that Darwin was influenced by his culture. The obvious version and the false version shift and meld so that neither can stand alone to be accused. Combining unoriginal truth and untrue originality creates an impression of original truth. So long as one accepts the poststructuralist assumption that all we can know is culture, these tactics are perfectly permissible. From that perspective, the closest one can get to truth is reminding people that truth is inaccessible; the highest moral purpose is making others experience the unreliable nature of human thought, and opening their minds to a wider range of perspectives. For those who do not share this poststructuralist view, Beer and Levine's rhetorical style makes their arguments at once impossible to build on and tedious to engage with. The only way to critique such arguments effectively is to point to the shifting tactics themselves.

The idea that all we can know is culture was attractive to humanists in the latter part of the twentieth century, and remains so, for a number of reasons. One reason is that it is partially true. Or, more specifically: part of the poststructuralist metaphysics of culture overlaps our actual situation in the human niche. As evolutionary anthropology and evolutionary psychology recognize, our social embeddedness is real. Humans are social learners, prone to elevate ourselves and our own cultures above others, even to distort what we see in accordance with our conventions and project our social structures onto the rest of the world. But that is not the whole truth. We are also asocial learners: dependent on interacting successfully with an unpredictable world outside our minds, evolved to perceive it well enough to survive in it. Our tendency to be moulded by our culture is constantly counteracted by our individual judgments, which are informed by our contact with other individuals and the external world, by conflicting conventions within our own culture, and often by contact with other cultures. Turning the half-truth of social embeddedness into the whole truth is a classic instance of projecting human social life onto the natural world—the same type of projection that went into the early evolutionary thinkers' treatment of natural selection as hero or villain. It is a move that plays seductively on our biases: our need to be center stage, our intense interest in social life, our sense of society's overwhelming power, and—through visions of individuals as equal nodes constructed by cultural webs—our egalitarian impulses.

For people specializing in the interpretation of culture, this overemphasis on our social embeddedness has an even more insidious allure. All scholars are more or less likely to exaggerate the importance of their subject, from sheer immersion if not from egocentrism. If we believe that all or even most of what we can know is cultural history, cultural analysis is invaluable. Rather than having to absorb information from other disciplines and take responsibility for its accuracy (recall Henkin's Piltdown predicament), humanist scholars can treat other disciplines as cultural history for us to analyze. Scientists may have mastered certain predictable ways of interacting with the physical world, but we can show them that most of what they do is project their own cultural assumptions on the unknowable, and that the things that matter for humans take place in the realm of culture. From this perspective, scientific theories about the world are relevant to humanists only insofar as they mirror the culture in which they were produced. Furthermore, the comprehensive hard scholarship of Henkin and Stevenson is no longer necessary. As long as we have a certain cultivation in our subject area, the assumed impossibility of reaching stable truth or explanation licenses us to pick and choose our evidence, turn words ambiguously, and rest conclusions on the most precarious linguistic associations. Our most important job is to provide mind-expanding experiences for our readers. Using our linguistic fluency we make people "see that language does not hold still"; using our capacity for fine rhetorical distinction we demonstrate "possibilities and abundance" in the simplest statement. The insidious part of this arrangement is that it lets us play to our strengths and feel that we contribute to society without making us responsible for advancing knowledge about our subject. It encourages us to shape reality not just to the human mind generally, but to our particular dispositions as humanist scholars.

Unlike the scientific method, the analytic method adopted by Beer and Levine does not provide much protection against cognitive, cultural, and personal biases. When they begin with astute observations, their view of culture leads them into surreal abstract arguments, and their rhetorical style lets them reintroduce human interest in the simplest possible way: by splitting their subject into opposites, by turning the opposites into hero and villain, and by seeing their own reflection in the hero. For instance,

Beer registers that Darwinian evolution is a unique threat to the human imagination. In line with Stevenson's analyses and my own argument, she observes that there has been a tendency to "domesticate" evolutionary theory—"to colonise it with human meaning, to bring man back to the centre of its intent" (2009 [1983], 7). But since she considers both evolution and the human imagination primarily products of culture, she must portray this clash as primarily a clash between cultural forces. She claims that Darwin was working "against the grain" of his cultural inheritance, but that he was also creating his theory from cultural tropes: "elements from older orders and particularly from recurrent mythic themes such as transformation and metamorphosis" (3, 7). When creating his theory, as she has it, Darwin "sought to appropriate and recast inherited mythologies" (3). This argument creates a rhetorical problem: If Darwin only challenged us to believe a slightly different mythological story, the conflict would not be very interesting. There simply would not be much at stake. But having already split her subject into old and new cultural clusters, Beer can introduce a battle between good and evil. The villain is what Darwin was supposedly working against: the "older orders" of natural theology and Aristotelian essentialism, representing naive subjection to "the categorizing zeal of human beings" (7, xxx). The hero is Darwin's "unsteadying of plot," the capacity of his theory to "rebuff common sense," and "overturn the observable world" so that we are set free into an infinity of possibilities (1, xxx). As a natural sequel, Beer admits that evolution is true in the same breath as she portrays it as a prevision of her own poststructuralism. Darwin's theory in her version becomes a wild, irrational inspiration: it "outgoes his powers of observation and is not coextensive with his reasoning" (90). Cast in this role, Darwin "deconstructs any formulation which interprets the natural world as commensurate with man's understanding of it" (90). This representation of Darwin allows Beer to evoke his real joy in biodiversity and individual variation, and to suggest the counterintuitive elements of evolutionary theory that still challenge our imagination. The cost is that she sidelines the explanatory impulse that prompted Darwin to put his theory together in the first place, and the scientific logic that gave his challenge its force.

Levine acknowledges Darwin's debt to natural theology and uniformitarianism, recognizes that natural selection was built on ideas of "continuity and order," and rightly observes that Darwin had to break with a tradition that considered species immutable essences (1991 [1988],

7, 24–84, 142). But since he has branded all these traditions equally (un)true discourses, Levine has the same rhetorical problem as Beer. His solution is like hers. His villain is a compound of the "continuity and order" on which Darwin built and the essentialist ideas with which he broke: it is a delusory "essential reality" that language "merely named," linked to the assumptions of uniformitarianism and "nineteenth-century realism" (1991 [1988], 7, 142). This vision of false order is pitted against a liberating threat from "deep contradictions within the Darwinian project"—a threat that matches the "postnarrative" theories of Michel Foucault and Edward Said, which "emphasize discontinuity and seriality rather than slow and constant change" (7). Within the novels and scientific texts Levine discusses, these forces constantly battle each other. The battle is represented by internal contradiction and enemy phrases turned allies in spite of the author—be it Austen, Huxley, or Conrad. No text is ever *purely* of one camp, which allows for analysis of ambiguities and grants an aura of sophistication to the dichotomy. But the dichotomy is still too simple to contain reality. Levine ends up claiming that Darwin had to avoid "definition," because it implies "an essentialist view of the world" and is thus "incompatible with evolution by natural selection" (143).

Stigmatizing order as "essentialism" contradicts Levine's original insight that natural selection is a theory based on ordered complexity, revealing ordered complexity. Darwin absorbed species categories from natural history and morphology, systems of adaptation from natural theology, and gradualist principles from geology. He tested these categories and systems in order to refine them or discard them, but when he did, he created more adequate categories and systems in their place. By extrapolating from Lyell's laws of geological change and explaining the genealogy of species that Linnaeus could only glimpse, Darwin revealed a biological reality that is both continuous and ever-changing, both governed by chance and interacting in predictable patterns. These features of nature are hard to assimilate; but they are complex statistical realities, not contradictions. Darwin's theory does not have an evil shell of essentialism holding back a force of righteous relativism. Levine's theory and method, like Beer's, conspire to make him sidestep the difficult assimilation of natural selection in favor of the simple moral dichotomies that fit most comfortably in the human mind.

On a basic level, Beer and Levine do not account for their subject. They set out to analyze how Darwinian evolution interacted with Victorian literature. Whether they call it "influence" or a "two-way street" or the "absorption and testing" of "ideas and attitudes," they undertake to show two phenomena in contact with each other (Levine 1991 [1988], 3, 5). Yet, they nowhere give a succinct account of natural selection. In their arguments, Darwin's theory is only hinted at, either as a set of cultural tropes or as a prevision of poststructuralism. They pick concepts to represent evolution in a seemingly arbitrary way, mixing theological positions, ideas that predate Darwinism, concepts fundamental to human perception, theories on which Darwinism is predicated, and popular Victorian metaphors. Thus, they are forced into the awkward position of arguing that Darwin made authors write about "growth and transformation," "retrogression," "extinction," "the web of affinities" (Beer 2009 [1983], 97–114, 148–56), "the human subject," "observation," "uniformitarianism," "change and history," "blurring of boundaries," "connections—ecological and geneaological," "abundance," "denial of design and teleology," "mystery and order," and "chance" (Levine 1991 [1988], 14–20). Even without poststructuralist assumptions in the mix, this account would not allow for a workable delimitation of the subject or a consistent analysis of it. Indeed, both Beer and Levine exclude overt evolutionary myth-makers like H. G. Wells and give primary attention to authors who do not engage with evolution in their novels, such as George Eliot and Charles Dickens. Levine makes a special point of discussing authors "who were probably not 'influenced' by scientific writing" (1991 [1988], 3). Beer and Levine offer just enough connection to Darwinian evolution to give the impression that they are discussing its interaction with literature.

The success of Beer and Levine is probably partly due to their expert use of the poststructuralist repertory, with all it entails of shocking implications, linguistic complexities, and sophisticated associations. They both write lively, confident prose that is dense with specific cultural references, and they probably gave many of their readers a first imaginative taste of evolutionary theory. But their success is certainly also due to the fact that they suggest an explanation for their subject. Unlike Henkin and Stevenson, they establish a research program of sorts—a theoretical framework that seems to reveal fundamental causal principles and that can be applied to any particular novel or scientific text in the subject area. According to this research program, scientific and literary writers are equal vessels for cultural streams that intermingle and part according to the irrational play revealed by Darwin (and Derrida) as natural law. The prominence of natural selection in these streams comes from its clash with influential cultural traditions like natural theology and Aristotelian essentialism, from its revelation of indeterminacy, and from the contradictory interpretations made possible by the supposed ambiguity of Darwin's language. Cultural scholars adopting this perspective do not have to register the range of specific challenges Darwinian evolution poses to the human imagination, and they do not have to face those challenges themselves. Instead, they can rely on familiar ideas and reduce natural selection to the epistemological tail-eating of their own cultural theories: "If we are as the theory describes then we cannot trust the theory" (Levine 1991 [1988], 272).

In the decades since Beer and Levine published, no alternative explanations have been offered for the fascination with Darwinism in literature. This is not the result of fading interest in the subject. Apart from the reference collections and reception studies already mentioned, recent monographs have been published by John Glendening, John Holmes, Virginia Richter, Ian Duncan, and Devin Griffiths. There have been several anthologies, and there is a continual outpouring of articles. None of these contributions strays far from the line established by Beer and Levine. Some treat evolution as an excavation ground for cultural tropes and a prevision of poststructuralism (Glendening 2008; Griffiths 2019; Richter 2011); others follow the older humanist tradition of tracing ideas from science to literature or categorizing cultural history, with occasional genuflections toward contemporary cultural theory (Duncan 2019; Holmes 2009; McLean 2009; Page 2012; Ruddick 2009).

These studies give ample credit to Beer and Levine. The 2005 conference anthology *Unmapped Countries*, edited by Anne-Julia Zwierlein, cites them both and proclaims in its introduction that all contributors emphasize "the discursive construction of 'biological reality' in the nineteenth century" (2005, 7). In the 2011 anthology *The Evolution of Literature*, James and Saul distance themselves from scholars who treat science as a "dominant discourse," and they approve of scholars like Beer and Levine who accept "the fact of the autonomy of culture and literature" (11, 12). John Holmes, writing in 2009 about Darwinism in Victorian poetry, pays homage to the "pattern" set by "pioneering books by Gillian Beer and George Levine," where "science was very much part of culture, and culture was best understood in its historical context" (1). The 2014 anthology *Reflecting on Darwin* deplores recent "attacks" on "social constructivism with an essentialist agenda" and positions itself as an addition "to the discussions introduced by Beer (1983) and continued by Dawson (2007), Glendening (2007) and Schmitt (2009)" (Voigts, Schaff, and Pietrzak-Franger, 2, 5). Devin Griffiths' *Age of Analogy* (2019) allies itself to the "generation of literary scholars, including Gillian Beer, George Levine, and Sally Shuttleworth" who "flesh out the common ground between literature and science" (8). Ian Duncan's *Human Forms* (2019) continues the conversation started by "Gillian Beer's classic monograph *Darwin's Plot*" and Levine's account of Dickens's evolutionism (160, 131–32). Glendening's *Evolutionary Imagination* (2008) represents a seemingly more conciliatory stance toward the reality of science. He claims to distance himself from the theoretical assumptions of Beer and Levine, but nevertheless reproduces the effect of those assumptions. Though disapproving of Beer and Levine's "constructionist view of science" that situates "Darwinism in the realm of story-telling," his own project "overlaps poststructuralism" because he happens to focus on the "contingencies, uncertainties, and confusions" that Darwinism produced in literature (8). His disapproval thus amounts to the "bemused" gestures toward the reality of science made by Beer herself, and his methodology allows no clearer fix on the subject than hers did.

My own explanation of the Victorian myths of evolution clashes fundamentally with the tradition established by Beer and Levine. I consider Darwinian evolution a unique challenge to the human imagination precisely because it is a "dominant discourse" and a "privileged form of cultural formation": that is, a systematic explanation of our origin that gets closer to the truth than any self-authored mythology. I treat the human mind as a product of evolution, and use an evolutionary framework to explain both the capacity for art and the basic shapes of art. Since culture depends on biological brains, I do not treat it as "autonomous"; but in line with current theories of gene-culture coevolution and biocultural research, I recognize that culture is all-pervasive in human life and necessary for our survival. Thus, I do not, like Beer, question the "survival value" of memory "across generations" (2009 [1983], 243). I deal with literary authors who directly addressed evolution or were clearly influenced by it. The range of humorous and agonized responses to Darwinian evolution in literature is not, in my view, a matter of different cultural streams clashing, metaphors subtly transforming themselves, or language rebelling against its own meaning. It is the result of individual people in a particular cultural setting responding imaginatively to an almost unimaginable truth about the natural world.

3.2 The Position of Evolutionary Literary Theory

The relationship my work has to previous scholarship is a special case of the relationship evolutionary literary study has to the rest of literary study. Evolutionary literary scholars hold that culture is not an autonomous force creating human behavior, that categorizing cultural history is not enough to explain literature, and that literary scholars should adopt a scientifically informed perspective that acknowledges both biology and culture. This call for a new research program has earned evolutionary literary scholars some positive attention from scientists and the general public (Cohen 2010; Fromm 2006; Goldstein 2006; Kean 2011). Other literary scholars have been less welcoming. During the 1990s and the first few years of the twenty-first century, evolutionary literary theory was not discussed within literary study. Since then, there have been mostly hostile responses from literary scholars. Some critics can be roughly aligned with the traditional humanism of Henkin and Stevenson (Crews 2008; Goodheart 2008) while others are closer to the theoretical persuasion of Beer and Levine (Greenberg 2009; Kramnick 2011; Spolsky 2008).

The most common charge against evolutionary literary study is that it is "reductive," meaning that it applies crude explanatory categories that fail to account for the complexity of literature, and of the particularity of individual literary works. That form of reduction is indeed a vice. It would be in science as well as in literary study, since it means failing to recognize the true nature of one's subject. Though the accusation of crude generalization sticks to the lower level of evolutionary literary study as it does to the lower level of conventional literary schools, it does not touch the best interpretive work that has been produced by evolutionary literary scholars since the end of the '90s—work that combines explanatory categories from the evolutionary framework with traditional humanist methods of cultural contextualization, biographical study, tonal and thematic analysis, and evocative close readings (Carroll 2005, 2012a, 2013a; Clasen 2010, 2012, 2017; Easterlin 2000; Gottschall 2008; Saunders 2009, 2018). These interpretive works are mostly absent from the bibliographies of articles criticizing evolutionary literary study, and key critical arguments outright deny their existence (Kramnick 2011; Fletcher 2014).

The criticism against evolutionary literary study has tended to skirt the actual cause of conflict. Along with the charge of reductionism, there have been attempts to discredit both the evolutionary social sciences as science and the knowledge evolutionary literary scholars have of the evolutionary social sciences (Kramnick 2011; Fletcher 2014). The possibility of absorbing any reliable knowledge from other disciplines has been called into question (Spolsky 2008). It has been suggested with regrets that scientific truths are too general to be applied in literary study (Crews 2008). These objections look as if they are predicated on the idea that evolutionary literary study would be desirable if it could work. They all have a tinge of special pleading. Explanatory categories taken from other disciplines have been used in literary study since it first became recognizable as a discipline (Abrams 1997). From the time of the New Critics and early cultural historians to the categorizers and theorizers of the present, we have used general categories to understand the particular—concepts of historical periods and language change, of genre, of narrative structure, of motivations and emotional effects, of psychological development and the nature of bonds, of social and political interaction, and of identity (Abrams 1953; Bradley 1905; Brooks 1947; Castle 2013; Watt 1957; Van Ghent 1953). All of these categories can be fruitfully informed by the level of generality in the evolutionary social sciences. The literary scholars who decry "reduction" in evolutionary literary theory typically use generalizations about human behavior from psychoanalysis, Marxist theory, gender theory, Foucauldian social theory, deconstruction, or moral philosophy. Some of these theories—notably psychoanalysis and Marxism—have become more controversial in their original disciplines than the most controversial hypotheses used by evolutionary literary scholars.

What separates evolutionary literary theory from other literary theories is not reduction, the use of generalizations, interdisciplinarity, or appeals to controversial hypotheses. It is the idea that knowledge about literature is continuous with knowledge about human behavior in the social sciences and natural sciences. Evolutionary literary theorists recognize that there are different levels of complexity in knowledge, different focal points, different levels of analysis, and emergent structures that interact intricately. But they hold that human cultural products do not exist in a realm separate from the rest of human behavior. The same epistemological rules apply: component parts are needed to understand emergent structures like organisms or cultural traditions, the same biases get in our way, and the same controlling systems are required for producing reliable knowledge. This view of knowledge implies that literary scholars have a responsibility to absorb research from a number of scientific fields related to their subjects, and also to apply scientific standards to their own work: restrain speculation through theoretical frameworks that are ultimately built on empirical tests, and test their own hypotheses where possible.

The most influential critical publication on evolutionary literary study is Jonathan Kramnick's "Against Literary Darwinism" (2011). It was published in the flagship journal *Critical Inquiry* at a time when evolutionary literary study had received a lot of public attention—when, indeed, according to Kramnick, it "would seem to be all the rage" (316). The article has an impressive citation count, and Angus Fletcher recently declared that "such is the force of [Kramnick's] critique" that he inadvertently put a damper on all connections between science and literary study: the supposedly more well-behaved efforts to use cognitive psychology "also seem diminished" (Fletcher 2014, 450). Kramnick claims to "take seriously the central premise of the Darwinian program in literary studies," and to "argue against literary Darwinism" on that basis (316). However, the central premise he has in mind is not the idea that knowledge about literature is continuous with knowledge about human behavior in the social sciences and natural sciences. It is "the idea that creating and enjoying literary works is an adaptation" (317).

Kramnick's article is essentially a rhetorical misdirection. All his forceful critique is aimed at a narrow version of evolutionary psychology represented by the early work of John Tooby and Leda Cosmides. In that version of evolutionary psychology, the human mind is seen as a set of "modules" adapted for special purposes on the Pleistocene savannah (317–24). If that were the only limitation of his scope, he might have hit one or two visible evolutionary literary scholars who remain dedicated to this model of mind. However, he conflates the Pleistocene modular view of mind with every argument for the adaptive function of the arts, and attaches it to key evolutionary literary scholars who had at the time critiqued the modular view of mind for a decade (324–33; see also Carroll, 2012b). As Kramnick presents the case, evolutionary literary study is abortive if there is not a "literature module": a distinct innate "process" like vision or language, following a set pattern of development (343). If literature as a behavior were as inflexible as vision, and if it had evolved during the Pleistocene, it would of course follow that literature is quite "invariant across the species and across time" (346). Indeed, "at the level of the phenotype, 'sitting around a Pleistocene campfire listening to the storyteller's tale' needs to be identical to reading *Middlemarch*" (326–27). This idea stands in sharp contrast to the best work that has been done by evolutionary social scientists and absorbed by evolutionary literary scholars since around the beginning of the twenty-first century. Had the argument against the adaptive function of literature hit home, some evolutionary literary theorists might have had to modify that highly hypothetical part of their explanatory framework. But the "literature module" bears no resemblance to actual hypotheses about the adaptive function of the arts, which envision interacting sets of cognitive dispositions premised on ideas of flexible general intelligence and gene-culture coevolution.

Because Kramnick attributes the wrong central premise to evolutionary literary study, he gives a skewed account of every scholar he purports to critique. The bulk of his article is spent setting up the fanciful "literature module" and knocking it down. In passing, however, he mentions objections to evolutionary literary study that suggest a real conflict: evolutionary literary study considers literature "a repository of timeless themes" and it is not "open-ended and modest" (346, 347). Kramnick assumes that there can be no such thing as "'deep themes that fascinate us in fiction'": "Academic literary criticism of course recoils from this sort of thing and for good reason" (338). In a gesture that ought to ease Fletcher's anxiety, he concedes that the "sciences of mind" can "provide suggestive points of linking between literary studies and more empirically driven disciplines" (347), but that linking must not involve any universal aesthetic patterns or imaginative dispositions. Science may enter literary study as isolated concepts describing particular cases: "connections between certain features of mind and certain kinds of texts or forms" (347). Such supposedly formalistic work should be complemented with a perspective where "cultural artifacts like works of literature are part of the environment our minds build, respond to, and build again" (347). Science cannot tell us how our minds were shaped by deep history to respond in certain ways to stories and images, but only "what is done with minds at certain times and in certain places" (347). The sciences of mind have thus been reduced to describing entirely context-dependent mental responses. Biology can have little to teach literary study, but literary study is not limited in the same way: "The historical and formal methods of our discipline in either case would have something to contribute to understanding how the mind works" (347).

The special pleading evident in the criticism of evolutionary literary study probably suggests deeper objections based on professional habits, territorial interests, and temperament. People who have chosen to become literary scholars are not likely to resonate to quantitative methods or have much training in them. To scholars of a particularly humanistic bent, "science" might suggest everything opposed to the verbal play and impressionistic meditation they cherish—it might feel like the colors of literary experience being spread across the periodic table and Imagination being shackled to statistical themes. Applying scientific standards to the study of literature might seem to threaten the independence and free play of literature as an art form. Or it might seem to threaten the rights of the individual to experience a literary work in an idiosyncratic personal way. It might seem to threaten any attention paid to political and moral concerns, or the study of influential non-scientific theories of human behavior. What it actually threatens is the freedom of literary scholars, in their capacity as academics, to speculate unrestrictedly about how human beings work. That threat is real. It is the same type of threat that Lyell's *Principles of Geology* posed to the free play of imagination exercised by geologists on the subject of mountain formation. Almost two hundred years after Lyell's threat was successful, mountains are still as imposingly present, still the cause of sublime experiences for travellers, and still the subject of artistic imaginative play—but geologists know a great deal more about how they work. There is a closer link between literary scholars as human beings and literature as a human behavior than there is between geologists and the physical world, but it is as erroneous to suggest that scientific standards in literary study will limit the free creation and experience of literature.

The relationship between evolutionary literary scholars and non-evolutionary ones has lost none of its inherent conflict, but some scholars on both sides have recently moderated their tone. In *Critical Inquiry*, Kramnick's "Against Literary Darwinism" was followed by Angus Fletcher's "Another Literary Darwinism" (2014). Fletcher is no less hostile than Kramnick to the idea that literary study is continuous with the larger world of scientific knowledge, but he suggests that scientific knowledge is relevant to literary study. Though he engages in the usual special pleading and suggests that there are no nuanced Darwinian literary interpretations, he also urges literary scholars to avoid "reducing Darwinism to an object of cultural poetics" (452). His position is that scientific knowledge and "literature" can "partner"—as long as they each "support the autonomous character of the other so that natural selection upholds the open-ended pluralism of literature and literature promotes the adaptive potential of Darwin's theory" (452). This supposed middle ground for evolutionary literary study is actually a continuation of non-evolutionary literary study. As an equal partner, science is an "unprivileged discourse," treated as a subject for inspiration but not a source of information. Since the fields are to be "autonomous," scientific concepts may be adopted as tools only on the assumption that they lose their scientific aspect the moment they enter literary study—that they become merely "discursive concepts" to be applied along with concepts from conventional literary study. So-called cognitive approaches along these lines have gained ground, and are indeed cited favorably by Fletcher (Fletcher 2014, 450; Zunshine 2010). The most important thing in his vision of equal partnership is that literary study must not become subject to scientific standards—or, in Levine's terms, that intimating abundance must remain more important than getting it right. Nevertheless, "Another Literary Darwinism" represents an increased need for literary scholars to pay lip service to evolutionary theory. Even a highly equivocal call for "mutual cooperation" (469) is a world apart from Beer touting Paul Feyerabend's suspicion that.

> the only difference between poets and scientists is that the latter, having lost their sense of style, now try to comfort themselves with the pleasant fiction that they are following rules of a quite different kind which produce a much grander and more important result, namely, the Truth. (2009 [1983], 84)

In line with Fletcher's conciliatory tone toward scientific methodology, there has been a surge of empirical studies of literature. The vast majority are not guided by hypotheses lodged in an evolutionary framework, but the journal *Scientific Study of Literature* is open to such theoretical approaches. In recent years, there have also been tentative inclusions of evolutionary literary study in casebooks and textbooks of literary theory (Castle 2013, 278–83; Parvini 2012, 177; Richter 2018).

While literary study has retained its practical rejection but sometimes substantially moderated its tone, key evolutionary literary scholars have softened their voices. In 2013, Brian Boyd restated his conviction that literary analysis "should be compatible with an evolutionary understanding of human nature," but reassured his readers that this does not mean literature should "*only* be investigated by incorporating Darwin" (3). Boyd's demand for compatibility with evolutionary theory effectively means a new research program for literary study: it means new standards for the field and a constant flow of scientific information to uphold those standards. He also calls for literary scholars to formulate and test hypotheses. These are unpalatable ideas to many literary scholars, at the core of the conflict between evolutionary literary study and the rest of literary study. But Boyd interlaces them as heavily with honey as Fletcher does his status quo "partnership." In Boyd's formulation, it appears that evolutionary literary study will cause no real inconvenience to those literary scholars who know nothing of evolutionary science:

> Indeed it follows from evolutionary principles that, since we have evolved to be so uniquely social and cultural, we have been fashioned to comprehend others of our species very well and to be adept at assimilating our local culture, including its literature, without needing to know a theory developed only a century and a half ago. (2013, 3–4)

This picture puts minimal pressure on Boyd's literary scholar readers, pitching the full force of evolved wisdom that they get for free against some recent theory that would cost them a lot to absorb. But the picture might arouse suspicion even among those least acquainted with evolution. Claiming that evolution has made us expert literary analysts is a fallacy as great as Beer's claim that evolution reveals us to be trapped in entirely unreliable minds. Academic debates the world over are fuelled by the fact that we are both blessed with insights about the world and burdened heavily with biases. Boyd's formulation conveniently omits the biases. Very little is said of the practical measures required to make intuition compatible with the counterintuitive facts of evolutionary theory. On the subject of scientific hypotheses, the emphasis is similar. Readers are invited to linger in the creative production of ideas without the discomfort of selecting and testing those ideas: "there exists no scientific method for generating true ideas. Dreams, drink, debate, or diligence may all spark ideas that turn out to be valid" (4). The final impression is that literary study may go on as it has done, as long as it conforms to the knowledge of evolutionary theory, which it is likely to do most of the time through the intuition of literary scholars. In the same issue of *Scientific Study of Literature,* Jonathan Gottschall independently arrives at the same conclusion as Boyd: "investigation in the humanities should be constrained, disciplined, and inspired by knowledge from the sciences, but I don't think literary Darwinism is the only responsibly consilient approach to literary study" (2013, 16). Without a sense of how scientific information is to be responsibly assimilated to literary study, these calls for a scientifically informed pluralism ring hollow. In Boyd's words, "assessing benefits without costs makes no sense" (4).

Boyd's and Gottschall's articles might both have appeared under Fletcher's title "Another Literary Darwinism." But despite vigorous rhetoric to the contrary, the gist of the two camps' conciliatory writs are no closer than Kramnick's "Against Literary Darwinism" (2011) and Boyd's "Getting it All Wrong" (2006). It is hard to say where this dual conciliatory movement will lead. But for a scholar focusing on the initial reception of Darwinism in biology, it is tempting to draw historical parallels. Contemporary literary scholars feel a pressure of territorial encroachment from the sciences while trying to cope with the imaginative challenges of evolution. Like the nineteenth-century theologians who faced similar problems, they raise concerns about threatened morality and dangerous political implications, and they try to cordon off areas of human experience from science. Evolutionary literary theorists, in turn, feel the pressure of moral opprobrium and professional isolation. As the conflict hardens, both sides adjust their rhetoric to relieve the pressure of the other side's criticism. Traditional literary scholars claim to be scientific and evolutionists claim to be traditional literary scholars. The historical parallel can be illuminated by Henkin's shrewd description of the "three stages" in which novelists responded to the clash between evolution and religion:

> Denunciation of the scientific doctrine characterizes the first stage. Since religion is truth, the syllogism read, and Darwinism is contrary to religion, Darwinism is not truth—and must be anathematized. When the mounting weight of evidence tipped the scales and proved the intrinsic truth of evolution, men were ready to discard religion as worthless, truthless. The syllogism now cut the other way. Despair through loss of faith characterized the novel in this, the second stage. Having touched the two extremes in a wide sweeping arc, the pendulum now began to oscillate gently between, and the evolution-theology crisis entered its third and final stage. Reconciliation of the truths of religion with the truths of the Darwinian theory marks this final phase, the denouement and curtain of this significant Victorian drama. (1963 [1940], 76–77)

Characteristically, Henkin draws the curtain on the "Victorian drama" as if it had ceased definitively before his own mid-twentieth century. As he depicts it, the Victorians passed down a synthesis of evolution and religion, and we no longer have to feel the "fears of faith in the presence of a godless science" (11). But as I have argued in previous chapters, such syntheses were radically unstable. A generation after Henkin, Bowler traced the ongoing drama up through the twentieth century (1989, 2007). Syntheses between science and religion have forced a heavy compromise from either science or religion. Before Lyell and Darwin, the Bible was a widely accepted source of knowledge for educated people about the historical development of our planet and its creatures. That is no longer the case.

I believe there is a similar syllogism at work in literary study. Either knowledge about literature is continuous with scientific knowledge about human behavior or it is not. If it is not, scientific study of human behavior needs serious revision to acknowledge an autonomous force in human life that it cannot analyze. If it is, literary study needs serious revision to absorb the scientific study of human behavior. Thus, the phases of the Victorian drama play on, and we must all strut our parts. My own ideal is like that equivocally expressed by Boyd and Gottschall, and unequivocally expressed by Joseph Carroll in the same journal issue (2013b): A research program that provides literary scholars with a continuous stream of scientific information, so that they can replace their current generalizations about human behavior with ones that are up to date and constantly tested. Such a research program would, as Boyd suggests, have to include the testing of hypotheses through scientific methods. In order to give a full account of literature, it would also have to include humanist methods of cultural and historical contextualization, biographical study, in-depth interpretation, close reading, and analysis of style. If this ideal program were realized, it would impose a heavy cost on literary scholars—particularly on those invested in non-evolutionary literary theories. It is difficult and time consuming to absorb knowledge from other disciplines, to evaluate competing hypotheses, and to remain informed about constantly developing research. However, scholars capable of making the transition could bring powerful new tools and theories to the understanding of literature. They could make real contributions to the ever-developing scientific understanding of human beings, and particularly to the as yet underdeveloped understanding of artistic behavior. They could even bring greater reach and relevance to the imaginative and evaluative reflection that might be most congenial to their temperaments.

References

Abrams, M. H. 1953. *The Mirror and the Lamp: Romantic Theory and the Critical Tradition*. New York: Oxford University Press.

———. 1997. "The Transformation of English Studies: 1930–1995." *Daedalus* 126 (1): 105–31.

Beer, Gillian. 2009 [1983]. *Darwin's Plots: Evolutionary Narrative in Darwin, George Eliot and Nineteenth-Century Fiction*. Cambridge University Press.

Bowler, Peter J. 1989. *Evolution: The History of an Idea*. Berkeley; Los Angeles; London: University of California Press.

———. 2007. *Monkey Trials and Gorilla Sermons: Evolution and Christianity from Darwin to Intelligent Design*. Cambridge; London: Cambridge University Press.

Boyd, Brian. 2006. "Getting It All Wrong." *American Scholar* 76 (1): 156–158.

———. 2013. "What's Your Problem? And How Might We Deepen It?" *Scientific Study of Literature* 3 (1): 3–7. https://doi.org/10.1075/ssol.3.1.02boy.

Bradley, A. C. 1905. *Shakespearean Tragedy: Lectures on Hamlet, Othello, King Lear, Macbeth*. London: Macmillan.

Brooks, Cleanth. 1947. *The Well Wrought Urn: Studies in the Structure of Poetry*. New York: Reynal & Hitchcock.

Carroll, Joseph. 1995. *Evolution and Literary Theory*. Columbia: University of Missouri Press.

———. 2005. "Aestheticism, Homoeroticism, and Christian Guilt in *The Picture of Dorian Gray*." *Philosophy and Literature* 29 (2): 286–304. https://doi.org/10.1353/phl.2005.0018.

———. 2012a. "An Evolutionary Approach to Shakespeare's *King Lear*." In *Critical Insights: Family*, edited by John Knapp, 83–103. Ipswich, MA: EBSCO.

———. 2012b. "An Open Letter to Jonathan Kramnick." *Critical Inquiry* 38 (2): 405–410.

———. 2013a. "Correcting for *The Corrections*: A Darwinian Critique of a Foucauldian Novel." *Style* 47 (1): 87–118.

———. 2013b. "A Rationale for Evolutionary Studies of Literature." *Scientific Study of Literature* 3 (1): 8–15. https://doi.org/10.1075/ssol.3.1.03car.

Carroll, Joseph, Jonathan Gottschall, John Johnson, and Daniel J. Kruger. 2012. *Graphing Jane Austen: The Evolutionary Basis of Literary Meaning*. Cognitive Studies in Literature and Performance. New York: Palgrave Macmillan.

Castle, Gregory. 2013. *The Literary Theory Handbook*. Blackwell Literature Handbooks. 2nd ed. Wiley.
Clasen, Mathias. 2010. "Vampire Apocalypse: A Biocultural Critique of Richard Matheson's *I Am Legend*." *Philosophy and Literature* 34 (2): 313–28. https://doi.org/10.1353/phl.2010.0005.
———. 2012. "Attention, Predation, Counterintuition: Why Dracula Won't Die." *Style* 46 (3–4): 378–98.
———. 2017. *Why Horror Seduces*. Oxford; New York: Oxford University Press.
Cohen, Patricia. 2010. "The Next Big Thing in English: Knowing They Know That You Know." *New York Times*, April 1, 2010, C1.
Crews, Frederick. 2008. "Apriorism for Empiricists." *Style* 42 (2/3): 155–60.
Duncan, Ian. 2019. *Human Forms: The Novel in the Age of Evolution*. Princeton: Princeton University Press.
Easterlin, Nancy. 2000. "Psychoanalysis and 'The Discipline of Love'." *Philosophy and Literature* 24 (2): 261–79. https://doi.org/10.1353/phl.2000.0033.
Fletcher, Angus. 2014. "Another Literary Darwinism." *Critical Inquiry* 40 (2): 450–69. https://doi.org/10.1086/674126.
Fromm, Harold. 2006. "Reading with Selection in Mind." *Science* 311 (5761): 612–13. https://doi.org/10.1126/science.1123990.
Gibbons, Tom H. 1973. *Rooms in the Darwin Hotel: Studies in English Literary Criticism and Ideas, 1880–1920*. Nedlands: University of Western Australia Press.
Glendening, John. 2008. *The Evolutionary Imagination in Late-Victorian Novels: An Entangled Bank*. Aldershot: Ashgate.
Griffiths, Devin. 2019. *The Age of Analogy: Science and Literature Between the Darwins*. Baltimore: Johns Hopkins University Press.
Goldstein, Rebecca. 2006. "A Cross-Cultural Relationship." *Nature* 441 (7090): 160–160. https://doi.org/10.1038/441160a.
Goodheart, Eugene. 2008. "Do We Need Literary Darwinism?" *Style* 42 (2/3): 181–85.
Gottschall, Jonathan. 2008. *The Rape of Troy: Evolution, Violence, and the World of Homer*. Cambridge; New York: Cambridge University Press.
———. 2013. "Toward Consilience, Not Literary Darwinism." *Scientific Study of Literature* 3 (1): 16–18.
Greenberg, Jonathan. 2009. "Introduction: Darwinism in Literary Studies." *Twentieth-Century Literature* 55 (4): 423–44. https://doi.org/10.1215/0041462X-2009-1001.
Henkin, Leo. 1963 [1940]. *Darwinism in the English Novel, 1860–1910: The Impact of Evolution on Victorian Fiction*. New York: Russell & Russell.
Hillegas, Mark R. 1967. *The Future as Nightmare: H. G. Wells and the Anti-Utopians*. New York: Oxford University Press.

Holmes, John. 2009. *Darwin's Bards: British and American Poetry in the Age of Evolution*. Edinburg: Edinburg University Press.
Huxley, T. H. 2009 [1863]. *Evidence as to Man's Place in Nature*. Cambridge: Cambridge University Press.
James, Simon J., and Nicholas Saul. 2011. *The Evolution of Literature: Legacies of Darwin in European Cultures*. Amsterdam: Rodopi.
Kean, Sam. 2011. "Red in Tooth and Claw Among the Literati." *Science* 332 (6030): 654–56. https://doi.org/10.1126/science.332.6030.654.
Kramnick, Jonathan. 2011. "Against Literary Darwinism." *Critical Inquiry* 37 (2): 315–47. https://doi.org/10.1086/657295.
Levine, George. 1991 [1988]. *Darwin and the Novelists: Patterns of Science in Victorian Fiction*. Chicago: University of Chicago Press.
McLean, Steven. 2009. *The Early Fiction of H. G. Wells: Fantasies of Science*. Basingstoke: Palgrave Macmillan.
O'Hanlon, Redmond. 1984. *Joseph Conrad and Charles Darwin: The Influence of Scientific Thought on Conrad's Fiction*. Edinburgh: Salamander Press.
Page, Michael R. 2012. *The Literary Imagination from Erasmus Darwin to H. G. Wells: Science, Evolution, and Ecology*. Farnham: Ashgate.
Parvini, Neema. 2012. *Shakespeare and Contemporary Theory: New Historicism and Cultural Materialism*. New York: Arden Shakespeare.
Richter, David H., ed. 2018. *A Companion to Literary Theory*. Hoboken: Blackwell.
Richter, Virginia. 2011. *Literature After Darwin: Human Beasts in Western Fiction, 1859–1939*. Basingstoke: Palgrave Macmillan.
Ruddick, Nicholas. 2009. *The Fire in the Stone: Prehistoric Fiction from Charles Darwin to Jean M. Auel*. Middletown, CT: Wesleyan University Press.
Saunders, Judith P. 2009. *Reading Edith Wharton Through a Darwinian Lens: Evolutionary Biological Issues in Her Fiction*. Jefferson, NC: McFarland & Co.
———. 2018. *American Classics: Evolutionary Perspectives*. Boston: Academic Studies Press.
Spolsky, Ellen. 2008. "The Centrality of the Exceptional in Literary Study." *Style* 42 (2/3): 285–89.
Stevenson, Lionel. 1963 [1932]. *Darwin Among the Poets*. New York: Russell & Russell.
Van Ghent, Dorothy. 1953. *The English Novel, Form and Function*. New York: Rinehart.
Voigts, Eckart, Barbara Schaff, and Monika Pietrzak-Franger. 2014. *Reflecting on Darwin*. Ashgate Publishing Group.
Ward, Mrs. Humphry. 2013 [1888]. *Robert Elsmere*. Edited by Miriam E. Burstein. Brighton: Victorian Secrets.
Watt, Ian. 1957. *The Rise of the Novel: Studies in Defoe, Richardson, and Fielding*. Berkeley: University of California Press.

Zunshine, Lisa. 2010. *Introduction to Cognitive Cultural Studies.* Baltimore: Johns Hopkins University Press.
Zwierlein, Anne-Julia. 2005. *Unmapped Countries: Biological Visions in Nineteenth Century Literature and Culture.* London: Anthem Press.

CHAPTER 4

From Adventure to Utopia

When Huxley placed man in nature, it was a dual placement: the pickle-bottle of the Saturnian scientist and the plateau of culture elevated through self-conscious intelligence. The contrast between these two images is affecting enough, and it is played out in marvelous rhetorical color. But it is a contrast between just two self-images: one clinically detached and the other splendidly glorified. In the wake of relatively restrained scientists and philosophers, literary authors evoked the Darwinian human species across a vast emotional and intellectual spectrum. Not only were we shockingly like the specimens brought home by naturalists and reassuringly far risen through our accomplishments; we were ridiculous and grotesque, we were the finest product of nature, we were inherently vicious, naturally loving, doomed to tragic potentialities, and on the verge of utopian transcendence. We were half animal or fully animal or super-animal. Though the human self-image had been explored in these registers throughout history, the Darwinian context gave particular urgency and concreteness to the exploration. In a climate that made scientific authorities question the most fundamental imaginative self-image or affirm it by basing "Man's dignity upon his great toe," the field of self-definition was more than open (Huxley 2009 [1863], 109). It was an alien wilderness that needed to be navigated. The human imagination was under the unique pressure of a non-human origin story,

E. Jonsson, *The Early Evolutionary Imagination*, Cognitive Studies in Literature and Performance, https://doi.org/10.1007/978-3-030-82738-0_4

and our verbal self-portraits showed the pressure through intense and sometimes contorted expressions of tragedy, comedy, satire and horror. We could begin to navigate the wilderness by summoning visions of ourselves, like terrifying or beckoning ghosts.

The simplest novels that engaged with Darwinism generally dwelled the least on the nature of being human. Fairy tales and adventure stories with a primary purpose to entertain could not afford to express theological and intellectual nuance. The course of least resistance for the human imagination can be seen from our earliest recorded mythologies to our current mainstream media: clear-cut agonistic structures, concrete themes from human life, and big symbolic categories containing pleasurable ambiguities—not hierarchically organized abstractions that are carefully defined and stringently qualified. Huxley used our penchant for vivid simplicity when he dismissed his opponents as pedants fiddling with anatomical detail and celebrated the sensible child who sees human greatness in everyday life (Huxley 2009 [1863], 109–111). But in the eyes of Emma Darwin, Huxley too strayed regrettably from vivid simplicity: she wished he would spend his verbal talent on "a Book" on zoology instead of on his scientific monographs—because she wanted "something that people can read" (Darwin 1864).

Novels that the general public "can read" may not discuss the precise delimitation of the human species or meditate on the consequences of inherited instinct for morality and free will. Nevertheless, such novels communicate a wealth of self-images with different tones—direct and emotional because of their relative simplicity. Fairy tales and adventure stories call up monsters, heroes, and angels that can be relatively unfettered by their intellectual implications. Such stories have been the stuff of mythology throughout worldwide recorded history, from Gilgamesh, Beowulf and the Homeric crowd through the Bible and every known pantheon. In his playful novel *The Lost World*, Arthur Conan Doyle turns the Darwinian origin story into straightforward adventure.

4.1 Arthur Conan Doyle's *The Lost World* (1912)

The human self-image in *The Lost World* is akin to Huxley's glorification in tone, but with more bravado, a more exclusively masculine European perspective, and a more clearly evoked contrasting image. The story follows an expedition to an isolated South American plateau where dinosaurs survive. Circumstances predictably turn dire when the protagonists in the expedition are trapped on the plateau and find themselves battling horrifying ape men together with the Accala Indians who live there. The ape men are the demonic contrast to the celebration of humanity represented by the protagonists and the Accala Indians.

The Lost World has received relatively little scholarly attention. Henkin mentions it in passing among evolutionary novels taking "the form of romance," and Ruddick gives it brief credit as "one of the most exciting adventure tales ever written... not totally shallow" (Henkin 1963 [1940], 267; Ruddick, 36). Rosamund Dalziell calls it an "imperial adventure novel" and details the way it was inspired by the South American expedition of Everard im Thurn (2002, 131). Amy R. Wong identifies a tradition, represented by Michael Dirda, Ian Duncan and Ross G. Forman, which holds that "Doyle's noticeably hyperbolic—even comedic—use of adventure tropes perform an awareness that adventure fiction was fading with the fortunes of empire" (Wong 2015, 60). Wong herself argues that *The Lost World* renews adventure fiction by disrupting "literary and journalistic convention," thus pre-empting T. S. Eliot's "ethos of adventurous form" (2015, 65–66, 77). In my view, Henkin's initial classification is closest to the truth. Though Doyle plays with the expectations of Victorian adventure stories, it is the showy performance of a master rather than the subversion of a critical ideologue. *The Lost World* has a simple plot built into a simple tonal and thematic structure, taking a minimal cognitive toll to produce a satisfying emotional effect. The creator of Sherlock Holmes himself judged it "more a boy's book than any I have done" (quoted in Wong 2015, 67). Given how masterly the novel is within its range—as Ruddick acknowledges—I believe it deserves analysis for what it is.

The protagonists of *The Lost World* are a showcase of ideal traits for Victorian men: passionate youthful risk-taking, penetrating scientific curiosity, and battle-hardened, calculating bravery. The narrator is a young journalist, "ardent," "tenderhearted" and "in excellent condition" (Doyle 2012 [1912], 11, 208, 233). He joins the expedition in order to win the favor of a lady. His leader is the formidable Professor Challenger—a famously contentious scientist with an "enormous" head and a gaze "very clear, very critical, very masterful" (25). They are complemented by Professor Challenger's renowned scientific rival Professor Summerlee and the "world-famous" "sportsman and traveler" Lord John Roxton (73). Though the different temperaments clash comically throughout the story, the protagonists are united by an ethos of courageous gentlemanly exploration. Professor Challenger holds passionately that "truth is truth" and shares with his scientific rival "that highest type of bravery, the bravery of the scientific mind" which "upheld Darwin among the gauchos of the Argentine and Wallace among the head hunters of Malaya" (87, 109). The rough Lord John Roxton, who has waged a personal war against slave drivers, comments that "there are times" when you "must make a stand for human right and justice, or you never feel clean again" (81). The primeval plateau challenges their bravery at first sight, surrounded by sheer cliffs with the skewered skeleton of an American at its feet:

> And yet, as we looked up at that beautiful fringe of verdure only a few hundreds of feet above our heads, there was not one of us who could conceive the idea of returning to London until we had explored it to its depths. (133–134)

The protagonistic ethos of *The Lost World* allies it to typical Victorian adventure stories. As in H. Rider Haggard's hugely popular *Solomon's Mines* (1885) and *She* (1886), intellectually and physically exceptional men dare all to discover the secrets of exotic lands. Rudyard Kipling projects a similar ethos of fair play onto nature without engaging with Darwinism—for example in *The Jungle Book* (1894), where the tiger Shere Khan is despised by all for hunting domesticated animals and juveniles. Doyle's adventurers become a Darwinian self-image only when they are aligned with the indigenous Accala against the ape men.

Though the adventurers are trapped on the dinosaur-haunted plateau, their primary concern soon becomes battling the ape men and protecting the Accala, so that "upon this plateau the future must ever be for man" (2012 [1912], 257). Professor Challenger distinguishes this clash of species from the human battles faced by other adventurers:

> What, my friends, is the conquest of one nation by another? It is meaningless. Each produces the same result. But those fierce fights, when in the dawn of the ages the cave-dwellers held their own against the tiger folk, or the elephants first found that they had a master, those were the real conquests—the victories that count. By this strange turn of fate we have seen and helped to decide even such a contest. (257)

This perspective imbues the Challenger expedition with cosmic significance. Against the foil of the ape men, the adventure story ethos expands through deep time and becomes a vision of man's earned superiority over the rest of life. Thus, Doyle channels the awe of human evolutionary history while making it safely meritocratic—and safely tailored to the merits of Victorian gentleman explorers.

The ape men are an extremely simple but powerful antithetical self-image. They quickly outshine the dinosaurs as the novel's antagonists. When the narrator climbs a tree for his first survey of the plateau, a face protrudes at him from among the branches:

> It was a human face—or at least it was far more human than any monkey's that I have ever seen. It was long, whitish, and blotched with pimples, the nose flattened, and the lower jaw projecting, with a bristle of coarse whiskers round the chin. The eyes, which were under thick and heavy brows, were bestial and ferocious, and as it opened its mouth to snarl what sounded like a curse at me I observed that it had curved, sharp canine teeth. For an instant I read hatred and menace in the evil eyes. Then, as quick as a flash, came an expression of overpowering fear. There was a crash of broken boughs as it dived wildly down into the tangle of green. I caught a glimpse of a hairy body like that of a reddish pig, and then it was gone amid a swirl of leaves and branches. (190)

This is a pointedly repulsive creature—sickly pallid and blotched, clumsy and bristled, aggressive, and cowardly. The unpleasantness of the ape men's half human state is amplified by their disgusting physical and moral traits, but disgust also makes them easier to dismiss as inhuman. Throughout the story, they display an underhanded violence that is almost comically antithetical to the novel's gentlemanly ethos. They kidnap humans in their sleep and lie in ambush to strangle them from the trees. They rejoice in biting and tearing their victims to death, gloat at their suffering captives, and "make a proper ceremony" of forcing those captives off a cliff: the "game is to see whether they are merely dashed to pieces or whether they get skewered" on canes of bamboo down below (225). The most that can be said for these apes is that "music of any kind" makes them good-humored, and that they show a disciplined cunning "which the larger beasts would lack" (225, 243). Their malevolence and unloveliness is such that the human heroes seem to be facing demons rather than tiger folk and elephants.

The division between gentleman and ape man gets blurred on a few occasions, but never enough to affect the overarching thematic opposition. Sometimes, the protagonists' valor seems to bleed over into the ferocity of their opponents. Lord John Roxton describes the ape men's human sacrifices as "horrible," but also as "doocedly interestin'": "We were all fascinated to see them take the dive, even when we thought it would be our turn next on the spring-board" (226). On the battlefield, the narrator discovers that "there are strange red depths in the soul of the most commonplace man" (233). Though his eyes have moistened "many a time over the scream of a wounded hare," the ape men put him in a frenzy of "blood lust": he is "cheering and yelling with pure ferocity and joy of slaughter" (233–234). Professor Challenger's eyes similarly shine "with the lust for slaughter" (257). He, the leader of the expedition and supposed "highest product of modern civilization," also turns out to look markedly like the ape men's king (232). They have the "same short, broad figure, the same heavy shoulders, the same forward hang of the arms, the same bristling beard merging itself in the hairy chest" (232).

Local satire, however amusing, does not constitute subtext. Challenger and his ape twin have been made much of by scholars. According to Ruddick, Doyle escapes shallowness by being "alert to the ironic thematic possibilities opened up when civilized men return to the jungle, amusingly noting the survival of primeval traits in the civilized Edwardians" (2009, 36). Douglas Kerr sees a similar ironic overlap and turns it into a bold claim for "Conradian" anti-colonialism (2016, 7). Forman, yet more boldly, declares that "on a formal level" Doyle "rewrites Joseph Conrad's *Heart of Darkness* (1898) in a Latin American setting" (2010, 31). These claims defy credibility. There is no sustained overlap between civilization and ape, and no invitation for the reader to side with the ape men. Indeed, the joke on Challenger is the only example of ironic overlap provided by either Ruddick or Kerr (Ruddick 2009, 37; Kerr 2016, 7).

There is no moral ambiguity in the violent passion that the adventurers share with the ape men. In Doyle's vision, the adventurers' "joy of slaughter" is justified by the ape men's moral and physical ugliness. It is a righteous fury, whereas the ape men's violence is mere sadism. The ape men's aspirations to humanity are predominantly portrayed with disgust, or as grotesque comedy. Lord John Roxton catches the essence of this tone:

> I call them apes, but they carried sticks and stones in their hands and jabbered talk to each other, and ended up by tyin' our hands with creepers, so they are ahead of any beast that I have seen in my wanderin's. Ape-men—that's what they are—Missin' Links, and I wish they had stayed missin'. (Doyle 2012 [1912], 221)

The protagonists do their best to lose the Missin' Links a second time. At the end of the war, the grown males among the ape men have been killed and the rest have been reduced to "a servile race" (261). The narrator describes the horror of the actual extermination, observing that "it needed a robust faith in the end to justify such tragic means" (257). However, the focus quickly shifts to a positive view of the end: "for us, the victory brought much advantage"—"the annihilation of the ape-men" marked "the turning point of our fortune" (258, 260). Though Doyle's protagonists are in a literal sense "agents of genocide," as Kerr suggests, the moral weight of that fact leaves no mark on *The Lost World* (Kerr, 8). The last view of the ape men is an unpleasantly self-indulgent allusion on the narrator's part. He likens the survivors in their slave quarters to "a rude, raw, primeval version of the Jews in Babylon or the Israelites in Egypt" (261). Any poignancy this parallel might have had dissipates in satire. The image of a "primitive Ezekiel" does not graft onto the demons that were lustfully slaughtered a moment ago (261). Biblical tragedy silently dwarfs bestial "wailings," and ancient Jerusalem looms above "the departed glories of Ape Town"—the glories of "about a thousand huts of branches and leaves" that have appeared solely as the site of ceremonial mass murder (223, 261). This is no ambivalent Conradian compassion, but a parting shot of ridicule from a supposedly tender heart at a thoroughly unlovable enemy.

The Accala Indians who live on the plateau join the protagonists literally and metaphorically. Their qualities help give universal appearance to the ethos of gentlemanly exploration. Their shoulders, unlike those of the ape men, could sustain Biblical grandeur without absurdity. They are "down-faced" but "plucky little chaps" who remain stoic through suffering and death (224). In order to save their comrades, they march on the ape men with decision, competence, and courage. Apart from their moral qualities, they are beautiful: "little, clean-limbed, red fellows, whose skins [glow] like polished bronze in the strong sunlight" and whose faces are "hairless, well formed, and good-humored" (231, 240). In Professor Challenger's easy language of racial hierarchy, their type is not "a low one," "whether judged by cranial capacity, facial angle, or any other test" (241). (The tribe below the plateau, in contrast, are "an amiable but degraded race, with mental powers hardly superior to the average Londoner" [39]). The Accala, with all their qualities, are placed securely below the members of the expedition. Professor Challenger is most pleased with them when they collectively bow in gratitude: "their deportment in the presence of their superiors might be a lesson to some of our more advanced Europeans"—"strange how correct are the instincts of the natural man!" (249) All the same, these ideal natural men are the archetypal contrast to the ape men. Soon after the hideous face of the first ape man appears, the Accala reveal their presence: "discs of light in every direction, ruddy, clearly-defined patches," "fires which could only be lit by the hand of man" (203).

Like Huxley, Doyle bases the dignity of the human species on the prefrontal cortex instead of the big toe. When Professor Challenger stands next to the king of the ape men, the "marked difference" shows "only above the eyebrows, where the sloping forehead and low, curved skull of the ape-man [are] in sharp contrast to the broad brow and magnificent cranium of the European" (232). That magnificent cranium contains the capacity for the Victorian gentlemanly ethos—and for the cultural control of nature that makes possible the human life strategy that the Victorians share with the Accala. The narrator is reassured of the human cultural power when he first sees the fires of the Accala after his fright with the ape man. He gets further reassurance at the peak of another nightmare. When he is being pursued, alone and frantic, by "one of the great flesh-eating dinosaurs, the most terrible beasts which have ever walked this earth," he falls into one of the Accalas' traps (208). Thus snatched from the jaws of death, he meditates on human ingenuity:

> I remembered that Challenger had declared that man could not exist upon the plateau, since with his feeble weapons he could not hold his own against the monsters who roamed over it. But now it was clear enough how it could be done. In their narrow-mouthed caves the natives, whoever they might be, had refuges into which the huge saurians could not penetrate, while with their developed brains they were capable of setting such traps, covered with branches, across the paths which marked the run of the animals as would destroy them in spite of all their strength and activity. Man was always the master. (211)

When the Accala later suffer losses in direct confrontation with a predatory dinosaur, the narrator claims to modify his view: humans live on the plateau "upon tolerance" (262). But like the moments of dissolution between ape man and gentleman, this adventure thrill does not seriously disturb Doyle's simple vision. Though the Accala suffer losses, the immense predator succumbs to a rain of poison arrows. Other primeval beasts do not fare better. The Accala herd the vegetarian giants like cattle, and they fish enormous creatures out of the water. In *The Lost World*, humans may have to work and risk death to master nature, but they are its masters—even in competition with the most terrible beasts that ever walked the earth.

Against the Darwinian implications of insignificance, bestiality and transience, Doyle places a garishly confident species portrait channeling the power of adventure story. When the narrator first sees the Accala fires, he exclaims, "How gloriously my expedition was justified!" (203). Having arrived in a world stuck in prehistory, seen live dinosaurs and a hitherto unknown anthropoid ape, he has wonder to spare for a population of humans: "Here indeed was news for us to bear back with us to London!" (203) Imaginatively, this is the same news that Doyle bears back to London. From primeval time onward, in the wildest of lands, man is master. This news offers more than the "hour of joy" he promises in the preface. The right reader might—consciously or not—feel the force of imaginative relief released by the clash between heroes and monsters: "At last man was to be supreme and the man-beast to find forever his allotted place" (256).

Because the human self-image in *The Lost World* draws so heavily on the Victorian adventure story ideal, it has limited scope. The story is, as Doyle says in the preface, aimed at persons suspended between boyhood and manhood. The ethos is gentlemanly, the Accala are pointedly fellows, and the enemy are ape men. Females contribute to the battle only catalystically: the narrator's Gladys is "the inspirer of noble deeds" without whom "there would have been no narrative," but she ultimately reveals "the twin shadows of selfishness and of fickleness" by marrying a solicitor's clerk in the narrator's absence (8, 10, 304). Doyle's heroes will probably appeal most to those who identify with the European civilization lauded by Professor Challenger as the summit of our development. Nations and ethnicities outside that summit are mostly incidental: we never get to know Accala individuals; a couple of South Americans appear as deceitful villains, and a loyal African servant sets up camp below the plateau to offer support. Nevertheless, Doyle's human self-image is more than masculine Victorianism. To begin with, it gains scope by being contrasted to universal antagonists. Modern readers might baulk at the flippant way in which the ape men are annihilated, but readers of any culture are likely to be repelled by their predatory ferociousness, sharp canines, reddish eyes, sickly blotchiness, bristles, and underhanded unrepentant malice. The ideal represented by Challenger's expedition and the Accala is not exclusively Victorian. Their courage, comradeship, strategic organization, and stoic fortitude equal masculine martial ideals that can be found from *The Art of War* to *The Iliad* to Shakespeare to modern war movies (see for instance Gottschall 2008). That ideal only captures a sliver of the human niche, but it could well resonate with non-masculine readers. For readers with a taste for complex literature, genre might be the decisive limitation. In order to enjoy *The Lost World*, one needs to be carried along through unmixed emotions—horror, disgust, laughter, anxious suspense, joy, and triumph. One needs to abandon oneself to the pleasure of a perfectly polarized world where good is good and evil is evil. If one is not willing or able to let that happen, the emotional cues might seem like loud, isolated trumpet blasts. And without the emotional engagement, Doyle's human self-image has little to offer. Conceptually, it is an assertion of human superiority based on a contrast with unintelligent or malicious beasts.

4.2 Robert Louis Stevenson's *Dr. Jekyll and Mr. Hyde* (1886)

Robert Louis Stevenson positions the man-beast more uncomfortably than Doyle. In *The Strange Case of Dr. Jekyll and Mr .Hyde*, Dr. Jekyll's evil impulses made flesh—the embodied psychological tendencies of a civilized man—is unmistakably simian. But unlike *The Lost World*, Stevenson's novel stops short at a Darwinian monster. Stevenson does not make any overt references to human evolution; he does not have his characters reflect on our species in relation to other species, and his novel is not an allegory of our evolutionary history. *Jekyll and Hyde* is above all a morality tale, thematically affiliated with Christian stories of virtue and temptation or Greek myths of hubris and downfall. Most of the story follows the lawyer Utterson as he attempts to understand why his upright friend Dr. Jekyll has become entangled with the vicious character Mr. Hyde. Utterson gradually uncovers Hyde's crimes and Jekyll becomes ever more secluded. Finally, Hyde commits a murder and Jekyll disappears. Utterson helps break into Jekyll's laboratory, where they find Hyde dead with a written confession from Jekyll: the two men were the same person. Hyde was Jekyll's evil self, released when Jekyll concocted a potion to divide his evil impulses from his good impulses. The morality tale is complete. Jekyll was impatient with his imperfection, used human powers to try to perfect himself, and instead destroyed himself through his sinfulness. The novel's epigraph is an excerpt from one of Stevenson's poems (originally about his ties to Scotland) beginning "It's ill to loose the bands that God decreed to bind" (Stevenson 2006 [1886]). The idea of divine fate suffuses *Jekyll and Hyde* as surely as the spirit of adventure inhabits *The Lost World*.

Stevenson's famous dual man is indispensable to a discussion of early Darwinian myth-making. Hyde does not inhabit a fundamentally Darwinian world—in his world, transcendental moral forces are cosmological facts, not just facts of human psychology. Nevertheless, he is one of the most notable among the early Darwinian self-images: a simian homunculus strapped inside humanity's very soul. In "Pulvis et Umbra," published two years after *Jekyll and Hyde*, Stevenson envisions the human moral struggle in unequivocally evolutionary terms (2009 [1888]). He marvels that there is a desire for good at all in the human creature: "this ennobled lemur, this hair-crowned bubble of the dust, this inheritor of a few years of sorrow," who would "deny in vain his kinship with the original dust" (2009 [1888], 298, 299). Throughout this essay, Stevenson evokes "the miracle of reproduction, with its imperious desires and staggering consequences," and he locates humankind in a natural world dominated by the struggle for survival: "all these prey upon each other, lives tearing other lives in pieces, cramming them inside themselves, and by that summary process, growing fat" (292, 293). The earth itself is "loaded with predatory life, and more drenched with blood, both animal and vegetable, than ever mutinied ship" (293).

No great span separates Stevenson's striving ennobled lemur from his unhappy dual man. Despite its sombre naturalistic vision, "Pulvis et Umbra," like *Jekyll and Hyde*, ultimately subordinates amoral nature to an affirmation of divine justice. The essay eulogizes a life of perpetual moral struggle and ends on a note of Christian perseverance: "Let it be enough for faith, that the whole creation groans in mortal frailty, strives with unconquerable constancy: Surely not all in vain." (301). If Stevenson had been miraculously unaware that *Jekyll and Hyde* had evolutionary implications, the novel would still have fed the post-Darwinian struggle of the imagination. Hyde haunted London at a time when he could not have been innocent of Darwinian associations: in 1886, a decade after the *Descent* reignited the all-pervasive evolutionary debates of the 1860s. Adding the background of Stevenson's agonized evolutionary awareness, the inner beast emerges as a Darwinian ape deliberately transformed into a character of Christian mythology.

Like Doyle, Stevenson has a history of being considered too much of a popular writer for literary scholarship; but he has received ample scholarly attention from the second half of the twentieth century onward (Danahay 2005, 21–24). Scholars have reached relative consensus about *Jekyll and Hyde's* basic level of meaning: the repressed Jekyll releasing the catastrophically impulsive Hyde, watched by the drily normative Utterson. As Benjamin D. O'Dell notes, the novel is usually considered to be "about psychological doubling and atavism" (2012, 510). This basic level of meaning has been appropriated to the various higher-level thematic allegorizations of contemporary literary study. During the late twentieth century, *Jekyll and Hyde* was caught up in a general enthusiasm for Freudianism, notably used by Stephen Heath and Elaine Showalter in arguments about repressed Victorian sexuality (Danahay 2005, 22; Heath 1986; Showalter 1999). More recently, Jekyll's repression has been given all the colors of poststructuralist identity politics: for O'Dell, *Jekyll and Hyde* "centers on the problem of class maintenance"; for Ed Cohen, it concerns "fears about the failure of masculinity as a coherent subject position"; for Stephen D. Arata, it is about "the relation between professional interpretation and the construction of criminal deviance"; and for M. Kellen Williams it is about severing "the link between a referent and its representation" (O'Dell 2012, 520; Cohen 2003, 181; Arata 1995, 233; Williams 1996, 425–26). These scholars rightly converge on an impression that the novel is about threatened social norms. They are not wrong that normative Victorian society includes criminal law, heterosexual families, professional authors, realist literature and a patriarchal class system. In my view, they are wrong to claim that *Jekyll and Hyde* allegorizes any particular one of these features of Victorian society. If the novel is to be understood in context, the evil of Hyde cannot become a side-effect of modern preoccupations with the evils of class, patriarchy, heteronormativity, or literary realism.

Jekyll and Hyde has been the subject of some historical scholarship with less overt political impetus. For instance, Anne Stiles sees the novel as an illustration of Victorian dual-brain theories and surmises that "from a late-nineteenth century perspective, Jekyll represents the pinnacle of evolution, while Hyde approaches its nadir" (2006, 884). Though Stiles makes a compelling case that she has found one of Stevenson's influences, her Victorian theories are by themselves no more adequate to explain the novel's meaning than the political allegorizations of her contemporaries. To my mind, Irving S. Saposnik finds the right level of generality when he claims that *Jekyll and Hyde* is about "the inevitable conflict between natural urges and societal pressures" (1971, 729). However, I believe Saposnik attributes twentieth-century secularism to Stevenson when he argues that Hyde is "an amoral abstraction" rather than a mythological fiend (730). Both "Pulvis et Umbra" and *Jekyll and Hyde* suggest that Stevenson acknowledged no such thing as amoral natural impulse. For him, good and evil are at war "through all the grades of life": he sees "ordered polities and rigorous justice... the law of duty and the fact of individual sin" even in "the swarming ant" (2009 [1888], 299). Indeed, his moral perseverance is fueled by the idea that all of nature takes part in the struggle between good and evil: "God forbid it should be man that wearies in well-doing, that despairs of unrewarded effort, or utters the language of complaint" (2009 [1888], 301).

Dr. Jekyll and Mr. Hyde represents a transitional form of imagination—an attempt to graft the new Darwinian vision of a predatory simian creature onto the dualistic cosmology inherited from Christianity. In Jekyll's world, there is pure good and pure evil at war in the human soul, and human flesh is shaped by the skirmish. His research is in "transcendental" medicine (Stevenson 2006 [1886], 64). He learns that the "natural body" is "the mere aura and effulgence of certain of the powers" that make up the "spirit" (68). The powers shaping the body are the two poles of Stevenson's moral dichotomy: "those provinces of good and ill which divide and compound man's dual nature"—"the thorough and primitive duality of man" (66, 67). Good is near saintly self-suppression, "effort, virtue and control," "honest hopes," the "dryness of a life of study" and "self-denying toils" (70, 72, 78, 80). Evil is "self-indulgence," "leaping pulses and secret pleasures," having "every act and thought centered on self," and "drinking pleasure with bestial avidity from any degree of torture to another" (73–74, 78, 80). Though Jekyll speculates about a future when each soul will be understood as more than dual—as a "polity of multifarious, incongruous and independent denizens"—that vision is only mentioned once, vaguely and in passing

(67). The moral dichotomy, by contrast, is vivid, and it pervades the novel the way Stevenson saw it pervading nature.

Stevenson's spiritual moral vision clashes sharply with the moral vision implied by Darwin's theory. In Darwinian nature, human conceptions of good and evil are not forces that shape matter; they are only the social adaptations and varied cultural transmissions of one species. Because Stevenson does not dramatize the clash between spiritual morality and naturalism, and because he does not waver in the performance of his spiritual moral vision, his novel does not so much meet the moral challenge of Darwinism as proceed regardless of it. But his moral dichotomy itself has a double nature. On the one hand, the spiritual conception of good and evil on which it is predicated makes an artificial division between pleasure and repression that leaves out most of human life. On the other, that conception of good and evil—because it is shaped by human social adaptation—is optimally shaped to please the human imagination. Stevenson's Good and Evil tap into our conflicting impulses toward individual assertion and self-effacing cooperation.

Jekyll's moral dilemma follows an ancient religious tradition of pitting spirit against flesh. His choice between pleasure and discontented control implies that pleasure is inherently selfish and dissolute. He can either encourage his "spiritual side" through penitence and charity or feed "the animal inside" through undignified and monstrous impulse (81). His evil tendencies come to the fore when he is "merrily disposed," and they "growl for license" though he has "chained" them (72, 81). In alignment with this moral tradition, Stevenson portrays virtue as a negative force—predominantly one of restraint. But as Darwin observed five years earlier in *Descent of Man*, positive forces like impulses and pleasures are not necessarily selfish. Human life can include the joy of camaraderie and shared experience, the love for a parent or child, triumph at constructive achievements, romantic longing, delight in puzzle solving, creative flow—or the elevation of tragedy, the glum satisfaction of fortitude, and the righteous gratification of punishing misdeeds. Either such pleasures cannot make humans contented in Stevenson's vision, or Jekyll is peculiar in that they could not make him contented.

Jekyll represents himself as the archetypal man without much gainsay. The good and evil in him are divided by "a deeper trench than in the majority of men," but the majority of men have that dividing trench (66). Jekyll has "a certain impatient gaiety of disposition," but the gaiety is only "such as has made the happiness of many" (66). Jekyll hides his undignified pleasures because of an "imperious desire to carry [his] head high," to "wear a more than commonly grave countenance before the public" (66). But this prideful self-image is a textbook tragic flaw. From a naturalistic perspective, his greatest idiosyncrasy is that his evil twin is sadistic. Hyde takes pleasure in "any degree of torture to another" (74). The only pleasurable misdeed that is described in detail is his murder of an MP, when he mauls his victim "with a transport of glee... tasting delight from every blow" (79). This sadism is treated as a natural consequence of self-indulgence—a potential in every human soul. Before Jekyll lets Hyde loose, his gaiety is "the worst of [his] faults" and for his pleasures he would "scarce use a harder term" than "undignified" (66, 73). At the end of his misadventure, he has to exercise "a great effort as of gymnastics" to hold back the sadistic impulses of Hyde, which deform him as soon as he loses concentration or sleeps (85). This makes him a stranger representative of the human condition than his discontent with ordinary human pleasures. Most people do not need to restrain themselves to avoid hurting others for pleasure. But Hyde's sadism feeds into the conflict between self and community that gives cross-cultural imaginative power to Stevenson's moral dichotomy. Hyde is so purely an expression of self-assertion that he is gratified by anything that diminishes other people.

Jekyll's potion is a fantasy of escaping the conflict between individual dominance and egalitarianism that haunts every human social system. His research reveals that evil is essentially "centered on self" and good is essentially "self-denying" (73, 80). He believes that it is "the curse of mankind" that "these incongruous faggots" are bound together, "polar twins" struggling in the "agonized womb of consciousness" (68). His greatest wish is not to suppress or express either impulse, but to divide the two: "if each, I told myself, could but be housed in separate identities, life would be relieved of all that was unbearable" (67–68). He creates the evil alter ego accidentally by compounding his potion in a mood of ambition instead of "under the empire of generous or pious aspirations" (71). Because he is selfish at the moment of transformation, he releases his evil side instead of his good side. Still, he tries to make the accident serve as a separation of identities. Hyde can assert himself through sadistic pleasure without tarnishing the name of Jekyll. The result is that Jekyll comes under the sway of Hyde while Hyde becomes a hunted murderer, open to the most ancient egalitarian punishment of bullies and law-breakers: the righteous crowd. Let him but "peep out an instant, and the hands of all men would be raised to take and slay him" (80). Jekyll claims to have learned his lesson, but he does not quite let go of his "beloved daydream" (67). In his last confession, written after the murder, he dwells on the "leap of welcome" he felt when seeing Hyde's face in the mirror: "In my eyes it bore a livelier image of the spirit, it seemed more express and single, than the imperfect and divided countenance" (71). He remains convinced that his potion could have produced "an angel instead of a fiend" (71). His lamentations at the failure of his fantasy almost drown his lamentations at the horrors perpetrated by Hyde.

Hyde's role as the evil pole of the moral dichotomy trumps his presence as a character. He has no personality traits beyond cunning malice and no wishes beyond evil pleasures: "nothing lived in him but fear and hatred" (83). Unlike the ape men of *The Lost World*, he is not even part of a demonic brood promoting their own group. Hyde is neither human nor ape, but "a being inherently malign and villainous," motivated by "causeless hatreds" (73, 85). He is "pure evil," pure undignified pleasures enlarged to "the monstrous"—in fact, he is a "child of Hell," a "foul soul" with "Satan's signature" written on his face (16, 71, 73, 83). He is an arch demon, and Stevenson clothes him in simian traits the way early Christians clothed their demons in the names and attributes of Babylonian gods. Our common ancestry with the apes becomes the latest connotative root of evil. This connection between ape and Hell fits conveniently with the tradition of beastly flesh threatening ethereal souls. But it also makes *Jekyll and Hyde* an active part of the imaginative self-explorations of humanity after Darwin.

Hyde is more insidiously chilling than Doyle's ape men because he combines archetypal cues for moral and physical disgust with the outward trappings of civilization. Like the monsters from the lost world, he is "pale and dwarfish" and gives an "impression of deformity"; his hands are "thickly shaded with a smart growth of hair," and he is "without bowels of mercy," impelled by "a sort of murderous mixture of timidity and boldness" (13, 15, 16, 75). He is often described as an overt semi-simian: "hardly human… something troglodytic," prone to "ape-like fury," "ape-like tricks," "chattering to himself" and jumping "like a monkey" (16, 23, 35, 84, 86). Despite these characteristics, Hyde wears the clothes of a Victorian gentleman, carries out correspondence in disguised handwriting, and occupies lavishly furnished rooms for his shady activities. When he tramples a child in the street, he faces outraged passersby with "black, sneering coolness" and appeases them by cashing a check at the bank (5). When he beats a politician to death, he does it using a fine cane that Jekyll got from the spotless Utterson (23–24). Though those who meet him immediately recognize him as "a fellow that nobody could have to do with," he persists for a time within the bounds of civilized society (6). He is an antisocial growth at the heart of the world's first industrial metropolis, and he comes in the guise of our closest animal relatives.

The downfall of Hyde is almost as divided in import as the soul of Jekyll. The arch demon's death at once re-establishes social justice and ends the fantasy of unmixed social impulses. The pure anti-social force has been banished, but the self-assertion it represents has become permanently established as a threat to society: "the doom and burden" of the struggling polar twins is "bound forever on man's shoulders" (68). Jekyll's fantasy of separating those polar twins lures readers along and leaves behind a queasy web of implications. In a world where Stevenson's Good and Evil walk as separate entities, Good would likely fare no better than the MP who asks Hyde for directions: clubbed to the ground before their faces fairly change from "innocent and old-world kindliness" to "surprised and a trifle hurt" (23). Even in a world without Hydes, a creature of pure self-suppression—stripped of self-assertion and individual desires—would be unable to cope with natural dangers. Had Jekyll produced an angel as perfectly self-denying as his fiend was self-centered, that angel might have succumbed even sooner than Hyde.

The lawyer Utterson, the focal character throughout most of the novel, is closest to the good pole of Stevenson's moral dichotomy. Utterson has all the contented self-denying control that Jekyll lacks. He is "austere with himself," drinks only gin at home "to mortify a taste for vintages," and "though he enjoyed the theatre, had not crossed the doors of one in twenty years" (1). He is "in any extremity inclined to help rather than to reprove" and, though emotionally undemonstrative, shows "something eminently human" in "the acts of his life" (1). At the final crisis when Jekyll's door is to be broken down, Utterson reassures the servants that they will not be held responsible for disobeying their master ("my shoulders are broad enough to bear the blame" [49]). Despite all these qualities, Utterson's contentedness is shaken as soon as he hears of Hyde's connection to Jekyll. He tosses sleepless, "enslaved" to "the problem" (12). Even at Jekyll's fireside, in "the pleasantest room in London," he is haunted by "the face of Hyde," and feels a rare "nausea and a distaste of life" (17). When Utterson accompanies a police officer to Hyde's rooms after the murder, his general distaste of life re-emerges as a horror of society. He sees a panorama of social dysfunction: a "city in a nightmare," "slatternly passengers," and "muddy ways" (25). He is even afraid of the police officer: "conscious of some touch of that terror of the law and the law's officers, which may at times assail the most honest" (25). Utterson never gets to comment on the truth about Hyde. We do not even find out if he follows Jekyll's scientific friend Lanyon into shock-induced death after learning that truth. But throughout the novel, his emotional responses to the events are guides to the truth. When Hyde leaves Jekyll's broken cane by the corpse of an MP, Utterson feels the vulnerability of society and the pressure of collective punishment. It is also through Utterson's eyes that Jekyll's evil side appears "troglodytic" (16).

Utterson's observations and emotional responses are the best candidates for a normative perspective in the novel, but that perspective does not stabilize Stevenson's dual image of humanity. Utterson offers no heroic determination to make the best of conflicting impulses, no fortitude in the face of the real Jekyll and Hyde. He sees that reality only in half-conscious glimpses, puzzling over it, presenting himself as the "picture of disquietude" (15). For most readers, he could probably not have been an antidote to Jekyll's potion even if he had been allowed to pronounce judgment. The lion's share of the novel's imaginative power, like its final word, belongs to "that unhappy Henry Jekyll" (88). Utterson is upright and kind, but he is a pointed nonentity. He is not an average human and does not struggle to be an angel. He was simply born more angelic than other people, "sometimes wondering, almost with envy, at the high pressure of spirits involved in their misdeeds" (1). His "catholicity of good-nature" is further supported by a lack of intellectual adventurousness and curiosity (1–2). Despite pursuing Hyde to ease his mind, he generally approves a rule expressed by his kinsman: "the more it looks like Queer Street, the less I ask" (7). His is a character to get through personal life with unerring virtue, not one to guide the human species through a wilderness of lost identity.

As a post-Darwinian self-image, *Jekyll and Hyde* is the tonal opposite of Doyle's trumpeted triumphalism. Humanity's fate is to have sadistic apes struggling against our best qualities. Those apes cannot be defeated in war and driven into slave quarters on distant plateaus. They cannot be joyously slaughtered using "the last word of the gunsmith's art from St. James' Street and the Strand" (Doyle 2012 [1912], 255). They are inside us, wheedling to be unchained, as much ourselves as our better selves:

"even if I could be rightly said to be either, it was only because I was radically both" (Stevenson 2006 [1886], 67). If we try to cast off these monsters, they return, like Hyde, "with more unfamiliar and more awful pressure" (68). Jekyll's role in this vision of humanity is to offer a fantasy of escape that ends in catastrophe, but also to reveal the truth about our dual nature. His friend Lanyon is the only one allowed to respond to the revelation:

> "I have had a shock," he said, "and I shall never recover. It is a question of weeks. Well, life has been pleasant; I liked it; yes, sir, I used to like it. I sometimes think if we knew all, we should be more glad to get away." (35)

Like Jekyll's, Lanyon's perspective is despairing to the point of melodrama—resigned not to living with conflicting impulses, but to leaving behind a life that is "unbearable" (68). Their catastrophic responses to the truth mirror the vaguer impressions of Utterson. It seems the only hope for a human individual is to be born contentedly balanced between evil and good, assertiveness and communalism, animal and spirit. If we are born slightly weighted toward evil, we will be hunted both by society and our own conscience. If we are born slightly weighted toward good, we might end up like Jekyll and "so vast a majority of [his] fellows," overzealous for "the better part" and being "found wanting in the strength to keep it" (78).

Stevenson's novel, like Jekyll, is a peculiar mixture of the universal with the personally and culturally idiosyncratic. The moral dichotomy at the center of the novel pits social against anti-social impulses in a way that would likely elicit some response from any reader from any culture. At the same time, that dichotomy requires readers to forget everyday truths—that not all pleasure is selfish, that selfishness is not sadism, that purely selfish and selfless humans could not survive or coexist. Readers only mildly inclined to explore Queer Street may see Jekyll's case as mere titillation, using Utterson as a dry and high place from which to watch. Readers constitutionally compelled to question life might be drawn to Jekyll's catastrophic ambition. The revelations of that ambition will most likely be less satisfying for those who have, like Darwin, seen the "polity" of conflicting human impulses that Jekyll could only glimpse (67). It will also be most satisfying for those who can share or overlook the fundamentally spiritual world view. In Stevenson's imaginative universe, those blind to the soul's dominion over the body are (as Hyde sneers at Lanyon) "bound to the most narrow and material views" (64). But through its caricature of the human condition, the novel reaches beyond the limitations of its thematic structure and specific theological origin. Out of the antiquated frames of its morality tale crawls an image that has more immediacy in the post-Darwinian world than prideful Oedipus, curious Pandora, Heracles at the crossroads, or Eve reaching for the fruit. Henry Jekyll's smooth brow contorts. His hands grow knotty and hairy, he sinks into a crouch, his clothes droop on narrowing shoulders, and he raises his cane like a brutal club. He has "nothing human," and he is just a step away from you in the street (83). The ape men encourage us to attack and conquer; Hyde wards us off and sends us into our minds to contemplate our own worst impulses.

Doyle and Stevenson both produced nightmare images under the pressure of Darwinism—foils to the ordinary human virtues of their society. In Doyle's case, the virtues foregrounded are martial competence, decisiveness, courage, and camaraderie; in Stevenson's case, they are (less emphatically asserted) restraint, studiousness, tolerance and charity. Despite the significant difference in the imaginative worlds of these two authors, the thrust of their Darwinian self-images is similar: humanity can maintain its dignity by controlling the vicious and selfish impulses of our evolutionary heritage. The essentially aversive effect of these self-images comes through when the Challenger expedition surveys Ape Town suited up for human sacrifice, or when Jekyll emerges out of the murderer Hyde, kneeling in a remorse so intense that he sees his "life as a whole" before his eyes (Doyle 2012 [1912], 231–232; Stevenson 2006 [1886], 79–80). Hyde and the ape men are simple, memorable monsters using horror and disgust to close off avenues of human behavior. In contrast, the monsters in Edward Bulwer-Lytton's strange novel *The Coming Race* (1871) provide very little guidance. Their agonistic ambiguity suggests that Bulwer-Lytton set out on an adventure narrative and, along the way, inadvertently experienced a cosmological identity crisis.

4.3 Edward Bulwer-Lytton's *The Coming Race* (1871)

Bulwer-Lytton was one of the first novelists to engage with Darwinism. *The Coming Race* was published in the same year as the first edition of *Descent of Man*, and though it has faded almost entirely in the present, it was widely popular with its contemporary audience (Page 2012, 126). On the most basic level, the novel is an entry into the same lost world genre as Doyle's novel—a subgenre of adventure stories launched by H. Rider Haggard but prefigured by Anglophone writers from Thomas More to Jonathan Swift. Bulwer-Lytton's adventurer Tish finds his lost world through a mine in an unknown location. He descends into the depths in pursuit of mysterious lights and voices, and reveals a subterranean civilization called the Vril-ya: humans who have evolved along an alternative evolutionary track and gained control over a mysterious life force. The Vril-ya astonish Tish with their angelic appearance, statuesque with mechanical wings and shining gemstones for halos. They also astonish him with their readiness to exterminate anyone who opposes their social order. Tish explores their domains, is threatened by execution because he is courted by their females (with whom he is not allowed to breed), and escapes through the aid of one of the courting females. Within the frame of this simple plot, Bulwer Lytton weaves a tangled web that encompasses imaginary technology, Christian imagery, caricatures of scientific debates, pratfalls, ambivalent feminism, theodicy, potshots at democracy, and giant lizards. When the novel first appeared, the tangle was so attractive that it inspired a "Coming Race Bazaar" at the Albert Hall and a "Vril-ya Club" partaking of "sun tans, moon bathing, and deep breathing" in order to bring on "human transmutation" (Bloom 2013, 40).

The Coming Race has, in modern times, been overlooked by both scholars and general readers. In 2012, Michael R. Page complained that Bulwer-Lytton, though "one of the leading novelists of his age," had become "much neglected" and that *The Coming Race* in particular had "largely gone underappreciated" (126, 131). Since then, one notable article has appeared, in which Terra Walston Joseph surprisingly claims that *The Coming Race* has "attracted the attention of numerous scholars"—a critical history substantiated by three articles, one of which includes "A Forgotten Satire" in its title (Joseph 2015, 233; Nayder 2004; Judge 2009; Wagner 1965). In a comprehensive introduction, David Seed observes that "the limited number of critics addressing *The Coming Race* have tended to take the novel as an anti-utopia ridiculing the inhuman order of the Vril-ya" (xvi). This is a tendency with complications. One of the novel's earliest reviewers, indeed, places it among "satiric utopias" (Seed 2005, xiii). Joseph, in her recent assessment, holds that Bulwer-Lytton follows "the typical strain of satirical utopian fictions," though she considers his criticism of political reform "less than total"

(Joseph 2015, 233). However, Page claims that *The Coming Race* "interestingly straddles the divide between utopia and dystopian satire" (2012, 129). Henkin goes one step further and declares that "to Lytton who based his ideal society on the thesis that 'mortal happiness consists in the extinction of strife and competition,' the civilization of the Vril-ya was Utopian" (1963 [1940], 237).

These radically different interpretations are not the result of idiosyncratic scholars, but of an idiosyncratic author. Bulwer-Lytton's incoherent ambiguity has prevented any consensus on the basic level of meaning such as has been achieved about *Jekyll and Hyde*. To Seed and Joseph, *The Coming Race* satirizes Women's Rights and mannish Victorian women, while to Page and Brian Aldiss it is an "overlooked" feminist novel "consciously trying to imagine a future in which such power relations among the sexes" is inverted (Seed 2005, xxxi; Joseph 2015, 233; Page 2012 [quoting Aldiss], 130). To Henkin, *The Coming Race* is a scientific projection, "one of the earliest pieces of fiction to exploit the theory of evolution and predict the development of man into a 'calm, intellectual race'" (1963 [1940], 235). To Bruce Mazlish, it is instead "strongly anti-materialistic, anti-democratic, and anti-Darwinian," seeking "to nullify the irrational aggressive threats to [modern society] in the unconscious by embracing some of those very irrational and aggressive forces" (1993, 740–741, 743). Judge sees the novel as a dystopian satire meant to "expose" the fact that all human societies rely on "thoughtless habit" (2009, 137). In my view, Page's characterization of the novel's "interesting straddle" between utopia and dystopia appears the most adequate. From one perspective, the Vril-ya are "superior utopians," and from another "hellish totalitarians" (Page 2012, 130). But that observation is just the opening for more fundamental questions: What is the nature of the straddle between utopia and dystopia? How does that straddle work, structurally and psychologically, for author and readers? Why and to what extent is it interesting? By extension, why has the novel been read so very differently, and faded so very decisively from its initial visibility?

The agonistic ambiguity of *The Coming Race* emerges gradually and never becomes coherent. The Vril-ya seem to have been conceived as saleable monsters—a nightmare with some spice from current scientific and political concerns. Like the Martians of H. G. Wells more than twenty years later, they are a product of evolution that spells doom for humanity. The adventurer Tish describes them as an almost omnipotent "distinct species" with which humans cannot coexist, because we cannot adopt their "modes of life" or "modes of thought" (Bulwer-Lytton 2010 [1871], 105). However, unlike Wells's Martians, the Vril-ya are superior in morals as well as power: they "unite and harmonise into one system nearly all the objects which the various philosophers of the upper world have placed before human hopes as the ideals of a Utopian future" (102). Since humans cannot coexist with creatures embodying our social ideals, Tish concludes that we "are not trained or fitted to enjoy for long the very happiness of which we dream" (102). This thematic structure could have been consistent with the Vril-ya as monsters. It could have been shaped into either a satirical or a tragic depiction of utopianism—a portrait of the human species foolishly denying its nature or heroically approximating an impossible goal. *The Coming Race* lands squarely in neither of these positions.

As Bulwer-Lytton unrolls the interaction between Tish and the Vril-ya, his hand slips in disparate directions. Tish is a naive American patriot, which makes him a comedic butt throughout the novel. The Vril-ya ridicule him with ample support from their author. But Tish also emerges as a sensitive observer registering the Vril-ya's tragicomical relation to humanity, and it is his voice that reflects on the meaning of his experience at the end. The Vril-ya are depicted as frighteningly powerful and chillingly conformist creatures from the first, but they are also beautiful, kind, peace-loving and intelligent; and they inhabit a perfectly controlled society that is genuinely happy. Their positive qualities may initially have been intended to make the rejection of utopianism more affecting, but those qualities take on a life of their own as Tish dwells on them. At different stages, their society becomes a lustful power fantasy and a tyrannical antagonist, a dream of social harmony and a nightmare of uneventful conformity. Through these turns, *The Coming Race* becomes a tug-o-war between human social ideals and current human nature with neither side unambiguously protagonistic. Bulwer-Lytton does not stand back from the tug to depict it deliberately. Through his ambiguous relationship with all of the novel's main characters, he flits from side to side to pull—and sometimes seems to take on the part of the rope.

Tish is not a showcase of human heroisms like Doyle's Challenger expedition. Nor is he a colorless paragon of virtue like Utterson, or a volatile tragic figure like Jekyll. He is an awkwardly constructed, occasionally self-aware compound of human prejudices and passions, proceeding through his adventure with fluctuating dignity and honor. Bulwer-Lytton swathes him in ridicule for his boast of American democracy, his slapstick attempts to master Vril-ya flying technology, and his polite mortification at the amorous attentions of the bold Vril-ya female Zee. The Vril-ya's perspective on him is consistently condescending, ranging from nurturing affection to a serenely pronounced death sentence. Tish even emphasizes his own ridiculousness by limiting himself to the name provided by his hosts: *Tish* is "a polite and indeed a pet name, literally signifying a small barbarian" ("the children apply it endearingly to the tame species of Frog") (69). But whereas the Vril-ya are certain that Tish's people are inferior to them, the small barbarian serves for Bulwer-Lytton to obscure the hierarchy between the Vril-ya and humanity.

Making Tish a stereotypical American means that Bulwer-Lytton can wash his hands of him even as he brandishes him against the Vril-ya. Tish the action-oriented individualist, impelled by "a taste for travel and adventure," is the natural adversary of the Vril-ya's idle conformity (1). In his most alienated mode, Tish considers the Vril-ya an "indolent race of philosophers" whose lifestyle oppresses him "with a sense of dulness and monotony" (97, 102). He longs for "a change, even to winter, or storm, or darkness" (102). Bulwer Lytton tacitly admits to the same feeling when the Vril-ya observe that those unaccustomed to "the serene equality of immortals" are likely to "find the happiness of gods exceedingly dull" (50). Though Tish is not a hero, Bulwer-Lytton joins him in defying the Vril-ya to produce the equals of human heroes—"poets, philosophers, orators, generals" (48). The novel's final defense of humanity, when Tish achieves his highest level of reflective authority, rests on the fact that the "social system" of the Vril-ya "forbid them to produce those individual examples of human greatness which adorn the annals of the upper world" (104). Tish even strikes a note of tragedy when he imagines the fate of Americans at the hands of the Vril-ya: "my brave compatriots would show fight, and not a soul of them would be left in this life, to rally round the Stars and Stripes, at the end of a week" (105). But both that tragic note and Tish himself are circumscribed by the fact that Bulwer-Lytton does not rally round the Stars and Stripes. As a British baron once offered the crown of Greece, he is an heir to what Tish dismisses as "the antiquated and decaying institutions of Europe"—the Europe that produced most of the utopian ideas on which the Vril-ya are based (19).

Though Bulwer-Lytton does rally to Tish's individualist passion, his birth allies him to the Vril-ya, with their "suave but lofty good breeding" and their well-ordered cities associated with "that little state of Athens" (67, 68). The novel's most heavy-handed satire occurs when Tish boasts of "that glorious American Republic, in which Europe enviously seeks its model and tremblingly foresees its doom" (19). He describes how the American model promotes "tranquil happiness": by preferring "for the exercise of power and the acquisition of honours, the lowliest citizens in point of property, education, and character" (19). The Vril-ya respond with an aristocratic horror worthy of Bulwer-Lytton's peerage. They reveal that their word for democracy is "Koom-Posh," meaning "Hollow-Bosh": "the ascendancy of the most ignorant or hollow" (36). Their theory of history contradicts Tish's hopes for American predominance: democracy inevitably "degenerates from popular ignorance to that popular passion or ferocity which precedes its decease"—a phase comparable to the French Reign of Terror, which they call "the universal strife-rot" (36). Indeed, a Vril-ya child predicts that the American Republic will enter the universal strife-rot "in much less than another 100 years" (68). And when Tish learns that the Vril-ya share the underground with democratic industrial states, his patriotic dream becomes dwarfed into a funhouse reflection. The occupants of these democratic states, almost identical to Tish's compatriots, are referred to as "savages" and exterminated like vermin when they become troublesome (64–65).

To the shared eyes of Bulwer-Lytton and the Vril-ya, the novel's only human character seems to be like democracy itself: it is "bad enough, still it has brains, though at the back of its head, and is not without heart" (68). Tish is an embodiment of human emotions rebelling against rationally derived social perfection. Though he hopes passionately that the American Republic will expand and enlighten the world, he settles into a grim certainty that the Vril-ya will "destroy and replace our existent varieties of man" (105). Without a single argument against their view of democracy as barbarism, he remains "proud to say" that he is "a citizen of a Koom-Posh" (68). With his brains at the back of his head, he is associated with the "animal organs" in the "hinder cranial hemisphere" that the Vril-ya have supposedly lost—including "amativeness and combativeness" and "the animal love of offspring" (47). His passions are not just those of Jekyll, tending toward violent self-assertion, but those of Doyle's adventure story heroes, tending toward death-defying achievements and honorable bonds. The very "restless aspirations" that Tish sees behind our dreams of a perfectly harmonious society are what make it impossible for us to inhabit such a society (102). We cannot "reconcile our passions to the modes of thought among the Vril-ya" because those modes of thought require a complacent obedience that would be fatal in "a society of contest and struggle" (104, 105). There are no Vril-ya heroes because complete social harmony makes "dormant or annulled" the "motives that impel the energies and ambition of individuals" (104). Tish is at once an embodiment of those motives and an inadequate vessel for them.

Unlike Doyle's heroes, Tish gets deliberately mocked for the bad side of individual passion. When he fantasizes about ruling the Vril-ya, he does not imagine himself a champion of human values reinstating his beloved democracy or reviving high pursuits like moral reasoning, scientific cosmology, competitive achievements, and art. Instead, he dreams of becoming absolute monarch, reintroducing alcohol, guns, meat eating, hunting, and war for pure conquest—a boyish revelry culminating in the personal satisfaction of being "deemed a demigod" (97). This could have been a caricature of Doyle's explorers, stripped of all motivations but self-seeking hedonism. But the mockery is not a condemnation of the passions themselves. The fantasy arises from the romantic attentions of the Vril-ya ruler's daughter, right after Tish has been dispassionately sentenced to death at the end of a prolonged polite interaction with an authoritarian society that is holding him prisoner. Though he lashes out with the individual dominance that is the stuff of archetypal antagonists, he is never motivated purely by individual dominance. He never behaves like Hyde, or like the citizens in the Vril-ya's account of democracy's decline, when "the brain and the heart of the creatures disappear, and they become all jaws, claws, and belly" (68).

Tish distinguishes himself from adventure story protagonists like those of Doyle not only by being wrapped in satire, but by being genuinely bewildered by his experience. When his brain is most to the fore, he recognizes the Vril-ya's "perfection of social or political felicity" (60). He acknowledges that they fulfill "the dreams of our most sanguine philanthropists" but he still wants to avoid "any ignorant disparagement" of humanity in comparison (104). Though he recoils passionately from the authoritarianism and conformity of the Vril-ya, he remains ambivalent to the last about his "duty" to humanity and his failure to return the morally perfect love of Zee (113). When he boasts of human heroes, Zee responds that progress based on "the predominance of a few over the many" is "utterly opposed" to the Vril-ya's "calm of existence," and he mirrors her judgment at the end of his narrative (48–49, 104). Though he uses that judgment to suggest that the Vril-ya are an undesirable ideal, he openly declares only that they are an impossible ideal—impossible "because we have so combined, throughout the series of ages, the elements which compose human character" (102). The accidental development of the human character is set against the accidental development of the Vril-ya character, and Bulwer-Lytton's inarticulate sympathies most closely resemble an uneasy fatalism. In his vision, human passions are not clearly antagonistic, but they are in unavoidable conflict with the egalitarian harmony of the Vril-ya, which is itself not clearly antagonistic.

If the Vril-ya's guest had been an avatar of vicious selfishness like Hyde or the ape men, the novel would clearly have been a satire of human nature. Equally, if he had been a European aristocrat reasoning against the Vril-ya on moral grounds, the novel would have emerged as a caricature of social idealism. Even a tonally neutral everyman falling short of Vril-ya standards would have produced a more consistently elegiac view of human utopian ambitions. But Tish—silly but upright, naive but reflective, assertive but not without heart—makes the line between satire and earnestness almost impossible to trace. He offers no alluringly humanized version of Utopia and no particularly appealing representation of human nature. He simply remains sympathetic enough so that readers can partially inhabit his perspective, experiencing the mixed protagonistic and antagonistic qualities of the Vril-ya. A reader following his emotional trajectory can feel attracted to the Vril-ya and repelled by them, puzzle about those emotional responses along with him, contemplate the accidents of our evolutionary development, and draw individual conclusions about possible or desirable human social existences. Yet, the cues of ridiculousness with which Tish is surrounded make it possible to maintain a distance both from his conclusions and from the fully disturbing effect of the Vril-ya. After all, the Coming Race did not face the best that humanity has to offer.

The Vril-ya would have produced an ambivalent emotional effect without being filtered through Tish's chimeric perspective. Their immense power comes from channeling a life force called *vril,* which allows them to manipulate matter and minds alike. They fear nothing from nature or other civilizations. Adapting to their power, they have abolished war, crime, poverty, inequality, and carnivory, creating a social harmony so complete that "they are good (according to their notions of goodness) merely because they live" (24–26, 30, 41). But their thousand-year peace has meant a psychological change that nearly eliminates human behavioral flexibility. They treat their thoughts and emotions the same way as they treat the organisms with which they share the underground—taming, pruning, or exterminating them as soon as they appear to threaten the social harmony (28). Though intelligent, they show no interest in novelties and express their individuality only as trivial variation within the all-pervasive frame of custom (32, 62, 84). Through their extreme cultural control of the environment, they have turned their cultural system into the equivalent of inflexible instincts, but they have also created an environment perfectly suited for the satisfaction of those instincts. They have escaped both the universal struggle for existence and the particularly human agony of flexible consciousness, at the cost of their particularly human character. And it is only a matter of time before their success makes them run out of underground territory so that they "emerge into sunlight our inevitable destroyers" (113).

Through their dreadful power, the Vril-ya command interest in the same way as monsters like Hyde and the ape men: they are a seemingly unstoppable hostile dominating force, valuing its own well-being above all. Through their emotional stability, constructive effort, pragmatic compassion, and egalitarianism, they inspire sympathy in the same way as Utterson, the Challenger expedition, and Jekyll's better self. When the fascination of power and the evocation of sympathy melt together, the Vril-ya become a power fantasy not very different from Doyle's revelation that moral Man is master over brute nature. And when their ultimate valuation of social harmony translates into compulsory conformity, the Vril-ya summon chilling previsions of *1984* and *Brave New World*. Together, these features stimulate the positive and negative emotions attached to the two ancient factors of human social life respectively: the horror and pleasure of egalitarianism as well as the horror and pleasure of individual domination.

The Vril-ya's combination of triumphant and predatory domination with harmonious and oppressive egalitarianism make them a fundamentally ambiguous agonistic construct. In Tish's account, these four agonistic modes jostle against each other without producing more clarity than his own conflicting qualities. Should we fear the Vril-ya as the extreme ends of our social impulses or see them as a consummate balance between those impulses? Each agonistic mode is made vividly present through imagery and example, so that the Vril-ya are plausible—psychologically real—as both nightmare and ideal. The tone in which they are depicted is something like that of Tish's first description of their underground landscape: "so splendid... yet so sombre; so lovely, yet so awful" (14).

Like Tish, the Vril-ya are never a force of pure domination, but their hostile power is immediately apparent. Tish's first view of a Vril-ya face awakens "an indescribable awe and tremor" (9). Even while contemplating its beauty, he feels a stir of "that instinct of danger which the sight of a tiger or serpent arouses" (9) The fear is so strong that he falls to his knees shuddering, covering his eyes. His sense of danger is substantiated throughout the story. The Vril-ya achieved peace not on enlightened principle but because they can exterminate one another with a thought (25). Their marriages have reached "moral perfection" because the females are more powerful than the males—and the males, in turn, survive because a female who murdered her husband 2000 years ago was killed by a group of other females (implicitly establishing the punishment for future murderesses) (31, 32). Tish's hosts are pleasant to him as long as he remains within the bounds of their social norms, but they are prepared to eliminate him the moment he strays from those norms—though he strays only by being born with traits that could "adulterate the race" and unwittingly attracting Vril-ya females (82). Despite their vegetarianism, the Vril-ya are the essence of an apex predator: prey to no other species with the power to destroy them all. Though their expressions are serene, they evoke "conscious superior power, compassionate and inflexible as that of a judge who pronounces doom"—the mark of "a race fatal to our own" (108). The portrait of their hostile power culminates in the prediction that they will destroy all of humanity.

The Vril-ya have primal and intellectual power playing together under the silky skin of angels. They do not inspire dread through sickly beastliness and deformity, but through serenely beautiful faces that signal their difference from humanity: "void of the lines and shadows which care and sorrow, and passion and sin, leave upon the faces of men" (11). This dreadful beauty is reflected in their whole society. Their cities are perfectly controlled visions of harmony, displaying the finest traits of architecture from ancient human civilizations, fashioned from precious metals, lush with cultivated vegetation, bathed in fragrance and instrumentalized birdsong (9–11, 17). They are themselves associated with supernatural beings from Egyptian and Etruscan art—but above all, with Christian angels (9, 11, 77). Their snowy mechanical wings and opal halos give contemporary picquancy to an archetypally appealing aesthetic vision. Their beauty reaches through the mists of ancestral cultures into the universal pleasures of smooth regular features, tranquil natural scenes, shimmering vibrancy and orderly abundance (Dissanayake 2000; Dutton 2009).

If Bulwer-Lytton had produced traditional utopian creatures—purely calm and morally superior—their imaginative tug against humanity might have been more uneven. Pure social harmony might have inspired only the queasy unreality of a flattering painting, or the festering suspiciousness of golden promises. Because they are so pointedly dangerous and so openly opposed to humanity, the Vril-ya do not fade like Utterson next to the passions of less blessed creatures. They are a calamitous force capable of tearing through the very earth with a thought (111). The devastating wonders of the vril power, referred to by Tish in his extremity as "detestable and unhallowed powers," have more than an occasional touch of Hell (98). Yet, the Vril-ya female Zee rends the earth to save Tish, and she carries him through the "terrible chasm" on her wings, "brightly and steadfastly and swiftly as an angel" (112). The note of otherworldly beauty is struck simultaneously as the note of danger: the Vril-ya's power is impossible to separate from the social harmony that is their claim to moral superiority. The resulting atmosphere is one of consummate harmony brimming with balanced power: a sensory evocation of the "social state," "so holy a contrast to the contentions, the passions, the vices of the upper world" (102).

Bulwer-Lytton lingers in the Vril-ya social state, evoking it through images of human social joys fantastically expanded. The Vril-ya live long, healthy lives, filled with innocent amusements and ardent romantic love that is perfectly faithful (102). Each of their city states is like "a well-ordered family," where citizens are autonomous but equal (67, 102). Those who have the temperament to pursue riches freely provide luxuries for those who do not (86). Their belief in a second life is so strong that they do not fear death, and when a loved one is lost, the family celebrates with a "Birth Song" that expresses "no regret, no farewell, but rather a greeting to the new world" (89). Tish, witnessing one such funeral, notes the "calm smile" still lingering on the face of the deceased, and the "cheerful face" of the son who just lost his father (89). The aim of all Vril-ya society is happiness "not as the excitement of a moment, but as the prevailing condition of the entire existence; and regard for the happiness of each other is evinced by the exquisite amenity of their manners" (47). This is the justification for the Vril-ya's absolute power: a social order that produces more peace and happiness for its citizens than any other social order. Tish recognizes their superiority in this respect, and despite his vindictive fantasy, he takes pleasure in the power.

Since the Vril-ya's moral system is an amalgam of human social ideals, their power translates to a form of divine justice in human eyes. They represent human morality capable of shaping amoral nature in its image. As masters of the ecosystem, they decide whether other creatures should live or die depending on whether they consider them "harmless" or "obnoxious" (80). A creature is obnoxious if it threatens their society, but also if it is generally destructive. Though Tish sympathizes with the underground humans who are destroyed on this principle, he notes not without relish that the Vril-ya eliminate "vast reptiles," "gigantic winged creatures, half bird, half reptile," and "lesser wild animals, corresponding to our tigers or venomous serpents" (28). The Vril-ya, as herbivorous apex predators, proscribe the existence of large predators. They illustrate this principle for Tish when they invite him to hunt for a "krek": an insatiable giant reptile that is all jaws, claws and belly, equivalent to amoral nature as antagonist (69). This creature is the antithesis to the Vril-ya's benevolent ideals, and they destroy it with pleasure: "what suffering it must have inflicted while it lived!" (71). By eliminating egocentric dominance and creating orderly harmony, the Vril-ya scaffold their domination of the ecosystem with human moral legitimacy. They personify extreme power for the greater good, to an alternatingly gratifying and chilling effect.

Though the vril power oscillates between protagonistic and antagonistic qualities—between being a power fantasy and a threat—the moral superiority of the Vril-ya means that they do not become fully antagonistic simply by threatening humanity. Readers are not forced to imagine the Vril-ya incinerating Doyle's and Stevenson's human virtues, only the warmongering underground states and Tish's "brave compatriots" (105). Bulwer-Lytton musters limited grief for the American people who, in Tish's version of events, will be doing their best to dominate the rest of humanity until they are interrupted by incineration (19). Though Tish's fear gives the Vril-ya a predatory power that complicates their protagonistic status, they join forces with human morality to the absurd Edenic extent of eliminating predators from the ecosystem. The positive and negative side of their domination are both possible for a human mind to inhabit sympathetically. The same can be said about the positive side of their egalitarianism, which is a fantasy of human social harmony. It is much harder for a human mind to sympathize with the negative side of their egalitarianism: the nightmarish conformity that surfaces periodically within their social harmony.

Conformity is the Vril-ya's most purely antagonist trait. Whereas the omnipotence and the social harmony are extreme versions of desirable aspects of the human niche—control over the environment and prosocial interaction—the conformity is, like Utterson, purely self-suppressing. It is not made more psychologically appealing by the Vril-ya's moral superiority. Whatever the advantages of a tradition, compulsory conventionality strikes a blow at the heart of the human survival strategy: it would spell adaptive disaster for any human culture that was not actually omnipotent. The horror of conformity is unlike the horrors of simple adventure stories. It does not inspire negative feelings by summoning the ancient enemies of tribal coalitions that litter human mythologies—underhanded predators and bullies like Hyde, the ape men, or the krek. Instead, it summons a more recent danger: social systems so large and powerful that they can no longer be modified by individuals in adaptive relation to the environment. That danger is most at home in the mythologies of the twentieth and twenty-first century, from E. M. Forster's "The Machine Stops" (1909) through the great social nightmares of George Orwell, Aldous Huxley and Ray Bradbury, to the dystopian movies of the present. But though state-level societies and global culture are recent features in human evolutionary history, the conflict between individuals and egalitarian coalitions may be as ancient as our species—and it has always had two sides. Individuals have to make sacrifices to be part of society. Hyde's horror at the hands of all men resembles Jekyll's frustrated suppression of identity and Utterson's queasy alienation from London society.

In evolutionary terms, it is the Vril-ya's conformity that finally makes them inhuman. Their unchanging social systems have nearly eliminated human behavioral flexibility. Adapting to the vril has turned "custom" into "part and parcel" of their nature (84). They automatically prevent any thought or behavior that conflicts with the social order (19–20, 84). As a result, they have also come close to eliminating the scaffolding for human behavioral flexibility: imaginative culture. Because there is nothing new for the Vril-ya to tell or feel, no difficult decisions or significantly diverging life courses, there are "no new plays, indeed no imaginative works sufficiently important to survive their immediate day" (60). Their poetry is nearly "starved to death" and their painting, after losing the individual touch of artistic composition, has become like "heterogeneous fragments of a dream of art" (13, 61). They engage in scientific studies, but without aiming at discovery and explanation—only to improve daily comforts and provide topics of conversation (63). Lives within their society are so similar that they see no need to chronicle life stories (61). They refrain even from deep frowns and full laughter to protect the peace (61). Expressions of individuality appear only as trivial variation in the social pattern: different "creature comforts," "hobbies," and "predilections" (32). Their utopian happiness is achieved at the cost of the flexible cultural development that forms the core of the human niche.

Throughout the novel, Bulwer-Lytton simultaneously suggests that the Vril-ya have become inhuman and that we could have become them. Even more provokingly—that we could still become them. Until vril changed their conditions for survival and reproduction, the Vril-ya's political, philosophical and scientific history paralleled that of the upper-world humans (23–25) "They had their rich and their poor; they had orators and conquerors; they made war either for a domain or an idea" (24). Their movement away from familiar human nature was gradual but decisive, like clouds gathering for storm: the portraits in their historical gallery became "more tame and monotonous" "in proportion as the beauty and grandeur of the countenance itself became more fully developed" (54). Though Bulwer-Lytton's version of Darwinism is limited by a mysterious life force, speculations about "hereditary culture," and a sense that natural selection produces "superiority," the thrust remains Darwinian (21, 48). The basic idea behind the Vril-ya is that humans change physically through selective reproduction in adaptive relation to their environment. Our ancestors were something different from what we currently call human, and our descendants may be something as yet unimagined. Most challenging of all: the way we currently organize our societies may contribute to producing a future human species that we would hardly recognize. The Vril-ya female Zee hypothesizes that humans such as the nineteenth-century spiritualists could develop the vril nerve "in the course of one or two thousand years" (53). She also closes the passage to the upper world: "not to be re-opened by me, nor perhaps by others, for ages yet unguessed" (112). Would those ages unguessed be enough to change "human character" in accordance with vril? Certainly, this was the more or less serious hope of the Victorian Vril-ya Club.

By giving the Vril-ya an evolutionary past, Bulwer-Lytton changes the parameters of Utopia. *The Coming Race* is unlike future visions where government has brought out the best or worst in what humans currently are. The Vril-ya are neither purely fantastical creatures held up for contrast nor humans who have undergone purely social reform. They are recognized as a human species with a physical and psychological make different from our own. If readers side with them, they are not just objecting to current human social institutions, as they might if they side with the utopian society in William Morris's *News from Nowhere* (1890). Because the Vril-ya are openly inhuman in their adaptations to Utopia, siding with them means openly siding against the current characteristics of our species. Yet, their psychology is not an arbitrarily inhuman construct, but one shaped by human dreams of perfection. Whether our sympathies are with the Vril-ya or against them, they are a double challenge: to imagine the minds of creatures adapted to Utopia and to recognize how different our own minds would have to become to live there.

The Vril-ya are an intellectual challenge in a specifically Darwinian guise, and they showcase the imaginative range of our basic social impulses. By casting both egalitarianism and individual dominance in a double role as hero and villain—by evoking both tyranny and complete control, both harmony and complete conformity—Bulwer-Lytton achieves a remarkably conflicted human self-image. Viewed through the eyes of his contemporaries, the novel is comprehensible as a bestseller. It was a new type of utopia, challenging and alluring in its scope, with enough ambiguity to allow a range of responses: wistful identification with the Vril-ya as a moral and political ideal, nihilistic disgust at humanity in comparison, horrified recoil from them as symbols of blind idealism, or a combination of all. *The Coming Race* fed on popular genres like the lost world adventure story, the novel of ideas, and even the non-fiction travel journal. It tapped hot-button topics like theodicy, American democratic ambition, feminism, autocracy, communism, and evolution itself. But some of these very factors explain why it has faded. The story is presented in the rather clunky form of fictional anthropology, including chapters of exposition on the Vril-ya's language construction, religion and political organization (35–39, 40–43). It contains baubles like the caricature of democracy and the equally heavy-handed jollities about nineteenth-century evolutionary debates ("these two schools raged against each other; one asserting the [Vril-ya] to be the perfected type of the Frog; the other that the Frog was the highest development of the [Vril-ya]'") (56, 54–57). The conflicted human self-image itself, combined with the authorial distance from Tish, means that it is a narrative without a clear agonistic structure.

Agonistic ambiguity can be a mark of artistic sophistication. It is a challenge to the human mind in the same way as the amoral Darwinian origin story. It requires us to be suspended between human social impulses instead of being carried along by them; to attentively apply our own moral judgment, and to evaluate the application and impulses themselves. However, such activity requires cognitive effort that needs to be repaid in some way. Repayment may come through a resolution or revelation, through an authorial voice that becomes an ally in moral reflection, or even through the simulated sensory pleasures of style. Despite the reflective moments of Tish and the powerful evocations of the different aspects of the Vril-ya, *The Coming Race* does not repay its cognitive cost. There is no climax to the story, only a limited companionship with an author who hides behind double-edged sneers, and little to no stylistic satisfaction. Though the novel may provide some therapeutic release for readers sharing Bulwer-Lytton's ambivalence about human nature and social ideals that release ultimately peters out in incoherent ambiguity. The intellectual bravery represented by the Vril-ya—the audacity of depicting a fully inhuman human ideal—is crippled by the layers of satirical disavowal. In front of the grave evolved angel flits a human semi-buffoon whose name sounds like an exclamation of mild disapproval. Tish is like a suffix of "just kidding" attached to both sides of the novel's conflict.

Even if *The Coming Race* had been without infelicities of narrative and style, it would necessarily look less appealing through modern eyes. The Vril-ya's pragmatic autocracy wrapped in benevolent aims evokes images of genocide and oppression since experienced across the political spectrum. Indeed, Bulwer-Lytton's idea of a hidden species wielding an ultimate life force was absorbed by the Theosophists and transferred into the racially tinged occultism that fed Nazi Germany (Seed 2005, xlii–xliv). Against Tish's judgment, occult supremacist groups took the Vril-ya as an ideal, believing themselves to have "secret knowledge that would enable them to change their race and become the equals of men hidden in the bowels of the Earth" (Seed 2005, xliii). The Vril-ya's nonchalant pruning of their ecosystem also makes it harder for modern readers to accomplish the imaginative leap for which Bulwer-Lytton asks. Though Bulwer-Lytton cannot be blamed for ignorance of a field barely in its infancy, the subsequent development of ecological ideas has struck a hard blow against his novel of ideas. The increasing understanding of ecosystems since Darwin's day has put pressure on the genre of utopian novels itself. And though *The Coming Race* ultimately depicts Utopia as inimical to human habitation, it depends for its effect on readers entertaining the theoretical possibility of Utopia.

There are no utopian classics to rival *We, 1984*, *Brave New World* and *Fahrenheit 451*. However, the utopian thinking to which those novels respond has hardly abated. Utopian fantasies are difficult for human minds to occupy, but they are also, paradoxically, basic projections of human social categories onto the rest of nature. They are appealing distortions that cease to be appealing the moment they are fully imagined. Perfect harmony would not only require a level of conformity and homogeneity that precludes adaptation to a changing environment—it would require perfect control over the environment, so that there are no longer any conflicts resulting from natural catastrophes or uncertain resource acquisition. Essentially, it would require banishing natural selection through domestication. Even if one could imagine a power like vril accomplishing that goal, one would also have to imagine a knowledge of ecosystems more complete than anything to which we can currently approximate. Would you destroy the large prey fish after finishing the krek by the lake? Unless perfect domestication is to end in ecological catastrophe, it needs to be supported by perfect ecological understanding. And if omnipotence and omniscience were both achieved, that still would not save us from inhabiting ecosystems based on the consumption of organic resources. Domesticated life, however harmonious on the surface, also lives by devouring life.

Both the unpopularity of utopian novels and the persistence of utopian thought can be illuminated by modern biology. Perfect social accord is the logical extreme of our niche's egalitarian impulses and joy in social learning; a complete lack of external dangers is the logical extreme of our need to control the environment through culture. Those extremes of harmony are at the core of most utopian visions regardless of historical context and political color. They are fantasies rooted in the cooperation and culture that form cornerstones of human survival. They may in some instances inspire social reform that increases harmony in real life. But fantasies of perfect harmony are at odds with other human impulses, such as impulses toward individual dominance, mental and physical exploration, and asocial learning. A utopian society may itself be a protagonist, contrasted to an antagonistic society like the kingdoms of Thomas More's Europe, the Bloody Sunday London of *News from Nowhere* or the eighteenth-century Britain that disgusts Gulliver when he returns from the Houyhnhnms. But a society free from all conflict by definition lacks protagonists, antagonists, right, wrong, just victories, poetic losses, and obstacles to overcome. It is anti-agonistic. To the extent that a utopian novel lingers in the utopian society itself, it is almost certain to be imaginatively unsatisfying.

Human beings are attracted to Utopia, as Tish suggests, through "our dreams of perfectability, our restless aspirations towards a better, and higher, and calmer, sphere of being" (102). It is a desire for social harmony that is, ironically, so ambitious that the Vril-ya would reach for their vril staffs to eliminate it. But human beings do not recoil from full expressions of Utopia, as Zee supposes, because we simply long to "quarrel with each other" (50). Our minds are not shaped to desire conflict but to interact with the rest of the world. Though Tish is an inadequate representative of our species, he is probably right that "a thousand of the best and most philosophical of human beings" could not last more than a year as Vril-ya citizens (104). Envisioning perfect harmony poses a problem for the human mind greater even than envisioning amoral Darwinian nature. Though non-teleological accidents and agonistic ambiguity take a cognitive toll, they reflect a reality of uncertain choices, motives and outcomes for which the human mind is somewhat equipped. No mind on Earth, human or otherwise, has yet experienced a life free from conflict.

References

Arata, Stephen D. 1995. "The Sedulous Ape: Atavism, Professionalism, and Stevenson's 'Jekyll and Hyde.'" *Criticism* 37 (2): 233–59.

Bloom, Clive. 2013. *Victoria's Madmen: Revolution and Alienation*. Basingstoke: Palgrave Macmillan.

Bulwer Lytton, Edward. 2010 [1871]. *The Coming Race*. Milton Keynes: Aziloth Books.

Cohen, E. 2003. "Hyding the Subject?: The Antinomies of Masculinity. *The Strange Case of Dr. Jekyll and Mr. Hyde."NOVEL: A Forum on Fiction* 37 (1/2): 181–99.

Dalziell, Rosamund. 2002. "The Curious Case of Sir Everard im Thurn and Sir Arthur Conan Doyle: Exploration & the Imperial Adventure Novel, The Lost World." *English Literature in Transition, 1880–1920* 45 (2): 131–57.

Danahay, Martin A. 2005. "Introduction." In *The Strange Case of Dr. Jekyll and Mr. Hyde*, edited by Robert Louis Stevenson. Petersborough: Broadview Press.

Darwin, Charles. 1864. "Letter to T. H. Huxley on 5 November 1864." *Darwin Correspondence Database*. "Letter no. 4661." University of Cambridge. http://www.darwinproject.ac.uk/DCP-LETT-4661.

Dissanayake, Ellen. 2000. *Art and Intimacy: How the Arts Began*. Seattle: University of Washington Press.
Doyle, Arthur Conan. 2012 [1912]. *The Lost World (Classic Reprint)*. Forgotten Books.
Dutton, Denis. 2009. *The Art Instinct: Beauty, Pleasure, & Human Evolution*. New York: Bloomsbury Press.
Forman, Ross G. 2010. "Room for Romance: Playing with Adventure in Arthur Conan Doyle's The Lost World." *Genre* 43 (1–2): 27–59. https://doi.org/10.1215/00166928-43-1-2-27.
Gottschall, Jonathan. 2008. *The Rape of Troy: Evolution, Violence, and the World of Homer*. Cambridge; New York: Cambridge University Press.
Heath, Stephen. 1986. "Psychopathia Sexualis: Stevenson's Strange Case." *Critical Quarterly* 28 (1–2): 93–108. https://doi.org/10.1111/j.1467-8705.1986.tb00010.x.
Henkin, Leo J. 1963 [1940]. *Darwinism in the English Novel, 1860–1910: The Impact of Evolution on Victorian Fiction*. New York: Russell & Russell.
Huxley, T. H. 2009. [1863]. *Evidence as to Man's Place in Nature*. Cambridge: Cambridge University Press.
Joseph, Terra Walston. 2015. "Bulwer-Lytton's The Coming Race and an Anglo-Saxon 'Greater Britain.'" *Nineteenth-Century Contexts* 37 (3): 233–48. https://doi.org/10.1080/08905495.2015.1030839.
Judge, Jennifer. 2009. "The 'Seamy Side' of Human Perfectibility: Satire on Habit in Edward Bulwer-Lytton's The Coming Race." *Journal of Narrative Theory* 39 (2): 137–58.
Kerr, Douglas. 2016. "Conan Doyle's Challenger Tales and the End of the World." *English Literature in Transition 1880–1920* 59 (1): 3–24.
Mazlish, Bruce. 1993. "A Triptych: Freud's The Interpretation of Dreams, Rider Haggard's She, and Bulwer-Lytton's The Coming Race." *Comparative Studies in Society and History* 35 (4): 726–45. https://doi.org/10.1017/S0010417500018685.
Nayder, Lillian. 2004. "Bulwer Lytton and the Imperial Gothic: Defending the Empire in The Coming Race." In *The Subverting Vision of Bulwer Lytton: Bicentenary Reflections*, edited by Allan Conrad Christensen, 212–21. Newark: University of Delaware Press.
O'Dell, Benjamin D. 2012. "Character Crisis: Hegemonic Negotiations in Robert Louis Stevenson's Strange Case of Dr. Jekyll and Mr. Hyde." *Victorian Literature and Culture* 40 (2): 509–521. https://doi.org/10.1017/S1060150312000071.
Page, Michael R. 2012. *The Literary Imagination from Erasmus Darwin to H. G. Wells: Science, Evolution, and Ecology*. Farnham: Ashgate.

Ruddick, Nicholas. 2009. *The Fire in the Stone: Prehistoric Fiction from Charles Darwin to Jean M. Auel*. Middletown, Connecticut: Wesleyan University Press.

Saposnik, Irving S. 1971. "The Anatomy of Dr. Jekyll and Mr. Hyde." *Studies in English Literature, 1500–1900* 11 (4): 715–31. https://doi.org/10.2307/449833.

Seed, David. 2005. "Introduction." In *The Coming Race*, edited by Edward Bulwer Lytton, xiii–liii. Middletown: Wesleyan University Press.

Showalter, Elaine. 1999. "Dr. Jekyll's Closet." *The Haunted Mind: The Supernatural in Victorian Literature:* 67–88.

Stevenson, Robert Louis. 2006 [1886]. *The Strange Case of Dr. Jekyll and Mr. Hyde*. London: Penguin Books.

Stevenson, Robert Louis. "Pulvis et Umbra." 2009 [1888]. In *Across the Plains with Other Memories and Essays*, 92–4. New York: Cambridge University Press.

Stiles, Anne. 2006. "Robert Louis Stevenson's Jekyll and Hyde and the Double Brain." *Studies in English Literature 1500–1900* 46 (4): 879–900.

Williams, M. Kellen. 1996. "'Down with the Door, Poole': Designating Deviance in Stevenson's Strange Case of Dr. Jekyll and Mr. Hyde." *English Literature in Transition, 1880–1920* 39 (4): 412–29.

Wagner, Geoffrey. 1965. "A Forgotten Satire: Bulwer-Lytton's The Coming Race." *Nineteenth-Century Fiction* 19 (4): 379–85. https://doi.org/10.2307/2932877.

Wong, Amy R. 2015. "Arthur Conan Doyle's 'Great New Adventure Story': Journalism in The Lost World." *Studies in the Novel* 47 (1): 60–79. https://doi.org/10.1353/sdn.2015.0011.

CHAPTER 5

Jack London's Evolutionary Imagination

> "I believe that life is a mess," he answered promptly. "It is like yeast, a ferment, a thing that moves and may move for a minute, an hour, a year, or a hundred years, but that in the end will cease to move. The big eat the little that they may continue to move, the strong eat the weak that they may retain their strength. The lucky eat the most and move the longest, that is all. What do you make of those things?"
>
> —Jack London, *The Sea-Wolf*

Jack London combined an extraordinary popularity with a continuous and explicit preoccupation with Darwinism. During his life, his fiction thrived in the borderland between boys' adventures and esteemed artistry. At the height of his career, he was "the highest-paid writer in America," and by the mid-twentieth century, he had "long been the most popular American author in Russia"—supposedly appreciated for his "action and outdoor setting" rather than his ideology (Haley 2010, ix; Bruccoli 1964, vi). When, a year before his death, he communicated his "mad appreciation" to Joseph Conrad, he was saluted in return as "a true brother in letters" whose "personality and art" merited attention: "for the vehemence of your strength and the delicacy of your perception" (Conrad 2015, 336). Ambrose Bierce considered "the hewing out and setting up" of Wolf Larsen in *The Sea-Wolf* "enough for a man to do in one lifetime" (quoted in Bruccoli 1964, vii). During the Cold War, Alfred Kazin

E. Jonsson, *The Early Evolutionary Imagination*, Cognitive Studies in Literature and Performance, https://doi.org/10.1007/978-3-030-82738-0_5

declared London an incarnate "confusion of all the innocence and lust for power in his day" (quoted in Berliner 2008, 55). More recently, he has been called "a high-powered dialectician," "America's best claim to a 'world author,'" and "a consistent bully" who "bullied himself to death" (Petersen 2011, 75; Reesman 2010, 34; DeCaire 2003, 290). Though the critical estimation of London's work has undergone changes as violent as his own life experience, his imaginative impact is beyond question.

Among the clashes of London scholarship from his own day onward, one idea approaches consensus: his worldview was inconsistent (Harpham 1975; Mills 1995; Petersen 2011; Reesman 2010, 29). There is less consensus about the nature and meaning of his inconsistency, but debates invariably connect to his socialism and his naturalistically inflected individualism. Scholarly opinion has sometimes stressed one or the other of these positions, ranging "from a conviction that in his eyes man was a mere animal and life only a brawl, to the nearly opposite persuasion that his philosophy was ... a social-minded collectivism" (Mills 1995, 3). As Gordon Mills noted half a century ago, both of these scholarly views "have a basis in fact"—and, indeed, London himself "was too intelligent to have missed the contradictions" (1995, 3).

The pertinent question is how London's contradictory worldview came to be, and how it affected his fiction. Some scholars in the late twentieth-century answered this question by approaching London as "a cipher for the economic relations of his time"—variously a testament to "the seductive nature of the capitalist episteme," to its "political unconscious," or to "socialist consciousness" (Berliner 2008, 55–56). More recently, the same evaluative interpretations of London's opposing views have been expanded along with the literary critical canon of "race, class, and gender, including revisions of naturalism" and occasionally "a poststructuralist approach" (Reesman 2010, 31). Other scholars have sought explanation in particular cultural influences—from Darwin, Spencer, Haeckel, and Huxley to Marx, Nietzsche, Jules de Gaultier, Aesop and the Bible (Berkove 2004; Faulstick 2015; Harpham 1975; Link 2010, 2013; Petersen 2011; Pizer 1983). Though studies such as these illuminate London's historical context, they essentially treat him as a vessel through which cultural and ideological tendencies entered his fiction.

London's contradictory worldview—and his own reflections on it—are explicable as his personal response to the unique imaginative challenge of Darwinism. Throughout his life, he was an amalgam of extreme impulses: intense ambition, self-direction, and overbearing confidence matched with a readiness for compassion, extreme gregariousness, an abhorrence of cruelty to animals and humans, and a desire for large-scale justice (Haley 2010). As Kazin suggested, this made him resonate with personal intensity to the intellectual winds of his time: the struggle for existence, the cry for social justice, and the indifference of the cosmos. These winds were not mere historical idiosyncrasies, however. Like the Victorians, London was staggered by Darwin's cosmological raw material. He responded by shaping it for human comprehension: reintroducing the elements of mythology through images of cosmic strife and harmony, envisioning nature and culture as embodiments of those forces at war, and embedding his horror of indifferent nature in larger systems of meaning. Scholars interpreting his work have followed his example after their own fashion: emphasizing either his cooperative or competitive tendencies, designating one or the other as evil, finding single cultural sources for those tendencies, and avoiding his genuine cosmic horror.

London's Darwinism has sometimes been dismissed as shallow and sensationalistic, or as so obsolete that it may be considered a historical quirk (Bruccoli 1964; DeCaire 2003; Pizer 1983). It has sometimes essentially been designated Nietzscheanism, with a sprinkling of Spencer (Harpham 1975; Mills 1995; Petersen 2011). A few scholars have observed that London had a steady intellectual marriage with evolutionary theory (Berkove 2004; Link 2010, 2013). He first read Darwin and Spencer in high school, and brought the *Origin* on his Alaskan gold trail before he achieved literary success (Berkove 2004, 243). In letters dated 1900 and 1914—approximately the beginning and end of his explosive literary career—he professed a belief in Haeckel's materialistic monism (Link 2010, 158–59). Though he became increasingly fascinated with Nietzsche in later life, he had found the core of his Darwinism at the fount long before: the idea of a unified cosmos guided by amoral laws, harboring life that develops through natural selection. This idea remained the base of London's worldview throughout the ethical and ideological changes of his career. Unlike the flippant tropism of Doyle, the Christian transmutation of Stevenson, or the jumbled abstractions of Bulwer-Lytton, London's Darwinism was wrapped around the roots of his mind.

Perhaps no other author has evoked the cosmic horror of evolution with the single-minded intensity of Jack London. The struggle for existence runs feverishly through short stories like "To Build a Fire," "The Law of Life," "A Piece of Steak," "War," and "Love of Life." In *Before*

Adam, a modern man nightly experiences "ecstasies of fear" as he dreams the life of a Pleistocene ancestor (2016 [1907], 17). Like his nightmare-riddled protagonist, London was haunted by evolution and at the same time believed strongly in its explanatory power. A fictionalized account of his scientific awakening appears in *Martin Eden*: the protagonist is enraptured with Spencer, proceeds to "a thorough study of evolution," wants biological interpretation "from the laboratory and the test-tube and the vitalized inorganic right on up to the widest aesthetic and sociological generalizations"—and then, he descends into suicidal depression (1982 [1909], 656, 772). In the more directly autobiographical *John Barleycorn*, London describes the same episode of suicidal depression—a "long sickness" supposedly caused by "pursuing Truth too relentlessly": "I read too much positive science and lived too much positive life" (1982 [1913], 1065).

London was right to rank his life along with his reading as part of his pursuit of "Truth." Biographies of his forty years have titles like *The Lives of Jack London* and *Jack London and His Times* (Haley 2010; London 1939). His ample firsthand experiences shaped both his worldview and his fiction in fundamental ways, starting before he had considered either. He grew up on imperiled small-hold farms and was buried on the 1000-acre Beauty Ranch purchased with his literary fortune. In between, he experienced boyish factory work, hard labor on trains and ships, oyster piracy, vagrancy, high school, a stint in prison, a university semester, a futile gold hunt, depression, alcoholism—and a failed marriage based on a scientific premise that he had debated in public with a possible mistress (Haley 2010; London and Strunsky 1903). He muckraked, lectured, agitated, war-corresponded, and sailed the world. Several of his harsh experiences took place before he encountered evolutionary theory in high school. Perhaps the protagonist of *Before Adam* is not speaking on a purely literal level when he describes the revelations of such an encounter:

> I had no knowledge of evolution until I was at college, whereas in my childhood and youth I had already lived in my dreams all the details of that other, long-ago life. I will say, however, that these details were mixed and incoherent until I came to know the science of evolution. Evolution was the key. It gave the explanation, gave sanity to the pranks of this atavistic brain of mine that, modern and normal, harked back to a past so remote as to be contemporaneous with the raw beginnings of mankind. (2016 [1907], 13)

Martin Eden's encounter with naturalistic cosmology is similar. His experience has been rich, but unstructured. He has been "curious to know things, and whatever he acquired he had filed away in separate memory compartments in his brain," until the memory compartments are connected through a vast vision of interconnected natural laws: "thus, he unified the universe" (1982 [1909], 654).

Through his self-education, London articulated and elaborated the tendencies established by his personality in contact with life. He became a firm Darwinian and an ardent socialist, deeply troubled by the contradictions he saw between human justice and the natural world (Mills 1995). He conceived of his own relentless ambition as both a mission to save the downtrodden and a righteous progress of the superior specimen. His life experience, thus transformed, enters specifically into his fiction, from the Klondike tales to the "twisted and stunted" factory boy turned tramp in "The Apostate" (London 1986, 219). As Matthew Bruccoli observes, his "life reads like something he might have written, a circumstance which has led to the charge that he egotistically dramatized himself in his novels"—and "the charge, as far as it goes, is true" (1964, v). But London's life readily transformed into legends of naturalism. When he dramatized himself, he dramatized all the positive life and positive science he had seen—as well as the depression that made him want to close his eyes on it, and the cajoleries he went through to regain his vision. He translated the petty brutalities and decent acts of Yukon prospectors (or their dogs) into the human condition under indifferent stars. He used his stories to act out his physical and theoretical encounters with the amoral cosmos, and to create symbols leading back to human morality. His fiction functioned for him on the small scale the way the arts may have done immemorially for humanity: guiding motivation through emotionally saturated, conceptually structured visions of the world.

The extreme tensions in London's personality and cosmic vision made his fiction interesting, but they also enveloped it in confusions, ambiguity, and self-deception. Toward the end of his life, he remained preoccupied with two novels that he himself considered particularly misunderstood: *Martin Eden* and *The Sea-Wolf* (Bruccoli 1964, xx). Both of those novels are combined psychodramas and cosmic allegories written at the height of his career. Both of them have histories of inconclusive scholarship. And both of them can be illuminated by an evolutionary analysis of London's worldview: an analysis that takes into account the challenges Darwinism poses to the human mind, combining biographical and historical information with a conception of basic human motives and biases. In turn, London's intense response to the intellectual winds of his time makes him a particularly apt case study of the imaginative challenges of Darwinism.

5.1 The Split Cosmic Vision and the Handcuff to Life

Though Jack London professed to be a monist, his cosmic vision was not so much contradictory as split. One part was the realm of naturalism, the other was the realm of morality. In his mind and throughout his life, the two vied mythically for power. He was excited by the explanatory power of evolutionary theory and thrilled to the competitive power of life, but periodically yearned for social harmony and cosmic justice. He found relief in the same way as it was found by T. H. and Julian Huxley: by envisioning morality as a cultural force in hopeless, heroic struggle against nature. That cultural force took different shapes throughout his life, but it was always associated with the archetypically protagonistic traits of unselfishness and constructive imagination, edged with the culturally specific imagery of Christianity. By seeing himself as a warrior in service of morality, London could satisfy both his competitive spirit and his prosocial yearnings. This self-image might have provided stability if he had seen a naturalistic basis for his moral vision. However, his naturalistic vision merged amoral natural laws with egoistic isolation, and his moral vision merged social justice with cosmic justice. He had to retreat from his naturalistic vision in order to pursue his moral vision and vice versa. Though he never retreated wholly from either, he was left emphasizing one or the other, agonizingly navigating the relationship between them. His memoirs and essays provide detailed accounts of this navigation.

London's naturalistic vision aligns him with the early evolutionary thinkers who focused on competition in nature. He saw struggle as so endemic to reality that he projected it onto the level of astrophysics: "the cool tumult of interstellar conflict," "the onset of starry hosts," "the screaming of planets and the crash of systems" (1982 [1909], 831). His naturalism was a hybrid beast: partly the truth that human values do not apply across the cosmos, partly the falsehood that human morality is an illusion. He felt the challenge of a non-mythological cosmology as coldly and clearly as a blade to the throat. He called it the "awful privilege of reason": "Man, with his brain, can penetrate the intoxicating show of things and look upon the universe brazen with indifference toward him and his dreams" (1982 [1913], 1094). He perceived with full imaginative presence how the human mind detached itself, transformed into Huxley's Saturnian scientist, and classified the human animal in a cosmos too vast for comprehension. From that perspective, the loss of cosmic justice and cosmic intention was clear to him: the "baffling head-reaches after immortality" were "but the panics of souls frightened by the fear of death, and cursed with the thrice-cursed gift of imagination" (1982 [1913], 940). Humans were "puppets of chance," "fluttering their May-fly life-dance of an hour" (1982 [1913], 940). But in London's mind—as in the most pessimistic mode of Huxley's—an amoral universe meant an amoral humanity.

To the end of his life, London struggled with the idea that biology revealed justice and good as "life's healthiest illusions" and humans—including himself and those dear to him—as "miserable little egotisms" (1982 [1913], 939, 940). Though he could locally recognize that his heroes had a "gregarious need for companionship," he never generalized such needs as part of his naturalistic vision (1982 [1909], 887). The hero of *The Iron Heel*, his socialist dystopian novel, professes belief in a social evolution similar to Spencer's and Darwin's visions of ever-expanding cooperation: "Combination is stronger than competition ... and man has been achieving greater and greater combinations" (1982 [1907]-a, 407). But even in this utopian glimpse, the focus was on combat for the sake of deliberate self-interest rather than on social needs and moral emotions: man "combined and made war upon his carnivorous enemies" in "a thousand centuries long struggle" (1982 [1907]-a, 407). He merged the struggle for existence with the class struggle, but did not explain the sudden cessation of struggle that would occur when both culminated in "that flower of the ages, the Brotherhood of Man" (1982 [1907]-a, 321). For London, struggle was the essential truth of the here and now even if it was not the goal. Egoistic isolation and naturalistic cosmology became an alloy, and they formed an "iron collar of necessity welded about the neck of his soul" (1982 [1913], 939).

London's idea of natural egocentrism was no doubt fuelled by his assertive personality. It may have been elaborated by Spencer's most radical Libertarianism and Nietzsche's master–slave rhetoric, both of which were alluded to in his writing. More clearly, the idea merged with his experience of depression. London called his egocentric naturalistic vision a "soul-sickness, life-sickness," a view through "the jaundiced eye of a pessimistic German philosopher"—but he also called it the "white logic," "fatal intuitions of truth," and the "truer order of truth" (1982 [1913], 936, 940, 1094). In his view, he had become depressed because he had seen this truer order of truth. He had made "the ancient mistake of pursuing Truth too relentlessly" (1982 [1913], 1065). Because of Truth, he had lost faith in "Success," "Recognition," "Love," and "Money"; he had come to consider cultivated people appalling in "their unlovely mental mediocrity" and art and culture "ridiculous" in the "face of the iron facts of biology" (1982 [1913], 1064–65).

During his depression, London decided that it was too costly to pursue the naturalistic vision as he had done. Though he still "read the books," he "pursued Truth less relentlessly"; he refused to take "the big things" too "seriously" (1982 [1913], 1067). He fought the vision of egocentric naturalism not because he had decided that it was untrue, but because he had decided that it was unhealthy:

> What is good is true. And this is the order of truth, lesser though it be, that man must know and guide his actions by with unswerving certitude that it is absolute truth ... Countless men have glimpsed that other and truer order of truth and recoiled from it. Countless men have passed through the long sickness and lived to tell of it and deliberately to forget it to the end of their days. They lived. They realised life, for life is what they were. They did right. (1982 [1913], 1094)

London's mythologization of naturalistic cosmology was more conscious than that of most Victorians. He openly declared that he was departing from the truth in order to tolerate life and find his way through it. However, his position was based on a truth that was half falsehood. As he faced his naturalistic vision, he was passionately and plausibly urged away from it by his moral emotions. In the same way, because his character and life compelled him to confront realities, he was irked by averting his eyes from the reality of cosmic injustice.

The moral vision in conflict with London's naturalism was consistently associated with culture. From a young age, he claimed to have glimpsed the cultivated world through literature and associated it with "unselfishness of the spirit, clean and noble thinking," "all that was fine and noble and gracious, all that gave decency and dignity to life" (1999 [1906], 88). When he found the real cultivated world full of "monstrous selfishness and sodden materialism," it strengthened his division between reality and morality (1999 [1906], 94). He became a revolutionary socialist; he aimed to "topple" the society that had disappointed him and to create a "new habitation," where man could "progress upon something worthier than his stomach" (94). From that point on, the revolution was his "ethical romance" and "Holy Grail" (93). Though he claimed to be essentially focused on "the PEOPLE" in this endeavor, he admitted that they were for him an essentially imaginative idea—a "radiant, flashing vision"—and that his fellow warriors were the socialist intellectuals among them: "keen-flashing intellects and brilliant wits" (1982 [1913], 1066; 1999 [1906], 91). Political idealism had replaced the moral idealism initially associated with cultivated society, but they were both pursued through culture, and conceived as forces of culture in conflict with the natural world.

London's socialism was in direct combat with cosmic injustices: "FATE," "hardships" and "accidents," and the powers in society that left some people more vulnerable than others to such injustices (1982 [1903]-b, 1118). When he wrote up his muckraking experience of the East End of London, he set socialist reform openly against his naturalistic vision, and couched it in religious terms:

> The application of the Golden Rule determines that East London is an unfit place in which to live. Where you would not have your own babe live, and develop, and gather to itself knowledge of life and the things of life, is not a fit place for the babes of other men to live, and develop, and gather to themselves knowledge of life and the things of life. It is a simple thing, this Golden Rule, and all that is required. Political economy and the survival of the fittest can go hang if they say otherwise. (1982 [1903]-a, 125)

In London's warring naturalistic universe, empathy could only exist as a spiritual axiom. Ironically, he expressed that axiom with equal doses of moral and combative emotion—but he could only see the combative emotion as part of nature. He had been "born a fighter" and socialism was an "illusion" worth fighting for (1982 [1913], 1065, 1066). He had summarily reasoned his own moral emotions out of the material world. Beyond the iron collar of his naturalism, his "soul" projected an aura for those emotions to inhabit.

London's most intense commitment to socialist revolution coincided with his recovery from suicidal depression. He believed that he had saved his own life by throwing himself into the "fight for socialism" (1982

[1913], 1066). It is probably not true that he fell into the abyss solely because he pursued naturalistic cosmology and was "handcuffed to life" by the "one remaining illusion" of socialism (1065, 1066). His descent into depression is related to the failure of his first marriage (Haley, 183–206). He mentions in passing that his fellow socialists put him "in touch with great souls" who retained "warm faith in the human," and that he completed his recovery from depression through "love of woman" (1999 [1906], 91; 1982 [1913], 1066). The far-off Grail of revolution would no doubt have been less effective without the more direct satisfactions of camaraderie and love. Whatever the actual causal forces at work in London's depression and recovery, his own interpretation of those events did not credit direct social satisfactions. The basic human needs for companionship and romantic bonds were just some of the healthy illusions that gradually returned to him: the stuff of "souls" and "faith."

Suicide, on the other hand, was scaffolded by naturalistic Truth. In his depressive state, London searched "all the universe" and found "but one freedom" for a man: "the anticipating of the day of his death" (1982 [1913], 939). He contemplated suicide "coolly, as a Greek philosopher might" (1982 [1913], 1065). The idea mastered him intellectually and emotionally:

> So obsessed was I with the desire to die, that I feared I might commit the act in my sleep, and I was compelled to give my revolver away to others who were to lose it for me where my subconscious hand might not find it. (1982 [1913], 1066)

London's interpretation of this depressive episode became a touchstone for him. The threat of Truth and the grace of illusion were imaginative truth for him, even if we judge them a misinterpretation—and imaginative truth was enough to shape his life and mind. His faith in the revolution had given him a non-suicidal means to oppose the naturalistic universe. His socialist mission allowed him "nobleness of purpose and heroism of effort," making his life "sunshine and starshine, all fire and dew" (1999 [1906], 91). But because his moral vision was not intellectually rationalized, it was fragile.

Though London based his socialist credo more on empathy than on "logic" or "economic argument," he distrusted the certainty of emotions (1982 [1903]-b, 1120). He *knew* that the suffering of others was wrong, and he easily imagined "the walls of the Social pit" closing around himself as it had around others—but in order to articulate his moral self-image, he "opened the books" and strove to be less "unscientific" (1982 [1903]-b, 1120). In his most pessimistic mode, the Golden Rule was merged with the necessary lie that "what is good is true," and both were associated with the blind emotions of non-human animals:

> Take a dray horse. Through all the vicissitudes of its life, from first to last, somehow, in unguessably dim ways, it must believe that life is good; that the drudgery in harness is good; that death, no matter how blind-instinctively apprehended, is a dread giant; that life is beneficent and worth while; that, in the end, with fading life, it will not be knocked about and beaten and urged beyond its sprained and spavined best ... This dray horse, like all other horses, like all other animals, including man, is life-blinded and sense-struck. It will live, no matter what the price. The game of life is good, though all of life may be hurt, and though all lives lose the game in the end. This is the order of truth that obtains, not for the universe, but for the live things in it if they for a little space will endure ere they pass. This order of truth, no matter how erroneous it may be, is the sane and normal order of truth, the rational order of truth that life must believe in order to live. (1982 [1913], 1093–94)

If life is good and what is good is true, there is a moral imperative underlying human society. That imperative is true, even if life in human society appears so cheap that the children of the East End are "for all the world like tadpoles just turned frogs on the bottom of a dry pond" (1982 [1903]-a, 36). And because that moral imperative is a healthy lie—a rational but lesser order of truth—it needs to be asserted forcefully, with the assurance of faith. In this way, London translated his subjective experience of empathy first into the spiritual axiom, then into instinctive delusion, without recognizing it along the way as a human impulse that was part of "the universe." As he saw it, the two alternatives were to be healthily sense-struck into moral delusions or to be perversely awake in amoral despair.

Because of his personality, London was peculiarly torn apart by this divided cosmic vision. He describes his youth as a time of "rampant" and "joyous individualism"; he was "proud to be one of Nature's strong-armed noblemen" and had a disregard for the weak that was only "hammered out" through personal contact (1982 [1903]-b, 1117, 1118). But when he did come into personal contact with those in the "Social Pit," he listened to their life stories with compassion; he found himself "looking upon life from a new and totally different angle" (1982 [1903]-b, 1119). He claimed to have been reborn by this experience. Nevertheless, both aspects of his personality remained. He translated his compassion into a "great oath" of self-interest: "I shall climb out of the Pit" (1119). As he climbed, his moral vision expanded from a personal work ethic to sympathy with his Pit companions to a studied and oft-professed socialism—but his socialist fables often culminate in the heroic achievements of individuals (Berliner 2008; Haley 2010; London 1983 [1908]; 1986, 221–36, 319–42). Through his scientific awakening, London was suspended between a realm of animated stomachs that fulfilled none of his prosocial needs and a faith-based morality that demonized his competitive self-assertion.

London's double nature meant that he often missed his professed mark. As Mills puts it, his writing "sustained two desires: one, the desire for adventure, combat, power; the other, the desire for friendship, justice, and a serene intellectuality" (1995, 8). London made himself unhappy with his divided world, but he also made socialists unhappy with his rampant individualism and individualists unhappy with his rampant socialism (Berliner 2008; Bruccoli 1964, vii–viii; Haley 2010, ix–7; Harpham 1975; Mills 1995). It was difficult for either camp to accept his "Nietzschean supermen whose actions incline thematically toward socialism" (Berliner 2008, 54). Geoffrey Harpham places London's intellectual position in a tradition of "Superman Socialism," but also notes that many of his friends "were appalled by his inconsistencies" (Harpham 1975, 23). London navigated this difficulty self-consciously but ineffectually. In the fictionalized prologue to *The Iron Heel*, he stresses that its towering hero Ernest Everhard was "but one of a large number of heroes who, throughout the world, devoted their lives to the Revolution"—but, he immediately adds, "it must be conceded that he did unusual work, especially in his elaboration and interpretation of working-class philosophy" (1982 [1907]-a, 319). The novel itself proceeds to celebrate Ernest's exceptional intelligence, rhetorical power, physical strength, and historical vision, all while lingering in a gruesome heroic struggle toward the Brotherhood of Man.

London opened the books of others to define his moral self-image, but he wrote his own to express it. The tension of his cosmic split is most vividly apparent in his fiction. There, the naturalistic realm and the moral realm clash in canine or human shape, or become colorful symbols leading protagonists to salvation or perdition. Individualist passions assert themselves against a universe brazen with indifference, and moral emotions achieve imaginative victories that were never mirrored in London's reasoning. His memoirs, essays, and biographical information provide the factual likeness of a remarkable life teetering on the cusp of naturalistic cosmology. His fiction, in turn, fulfills the promise of the fictional editor of *The Iron Heel:* it communicates "the *feel*" of Jack London and his times (1982 [1907] a, 319). London was obsessively focused on the horrors of natural selection and boyishly prone to simplistic idealism, but he intuitively understood the power of verbal art. He knew that things could be illuminated by science and still be hard to "understand":

> History tells us that these things were, and biology and psychology tell us why they were; but history and biology and psychology do not make these things alive. We accept them as facts, but we are left without sympathetic comprehension of them. (1982 [1907]-a, 319–20)

In this formulation, local and intuitive, London captures the imaginative problem of naturalistic cosmology—and the solution he himself, like Doyle, Stevenson, and Bulwer-Lytton, pursued through the arts. London had been cut by "the thrice-cursed gift of imagination" when it revealed the indifferent universe, and he used the same gift to dress his wounds. The best of his fiction both strives for sympathetic comprehension of Darwinian life and makes alive the late-nineteenth-century climate of evolutionary myth-making.

Though London wrote for commercial success—and achieved it—his imaginative capacity gave him a peculiarly intense artistic ambition. That artistic ambition, like his scientific awakening, is plausibly outlined in the autobiographical novel *Martin Eden*. Martin is "extraordinarily receptive and responsive, while his imagination, pitched high" is "ever at work establishing relations of likeness and difference" (1982 [1909], 559–60). He has a powerful aesthetic responsiveness that merges with his life experience and shapes the vision of his life:

> He was a harp; all life that he had known and that was his consciousness was the strings; and the flood of music was a wind that poured against those strings and set them vibrating with memories and dreams. He did not merely feel. Sensation invested itself in form and color and radiance, and what his imagination dared, it objectified in some sublimated and magic

> way. Past, present, and future mingled; and he went on oscillating across the broad, warm world, through high adventure and noble deeds ... (576)

Like Martin, London needed to express life: "its foulness as well as its fairness, its greatness in spite of the slime that infested it"; "song and laughter, and success and pain, and struggle and death" (1982 [1909], 666, 669). That need is evident in the best of London's novels and short stories, but the need itself is nowhere as clearly portrayed—and "*felt*"—as in *Martin Eden*.

5.2 *Martin Eden* (1909)

Martin Eden is a fictionalized depiction of Jack London's rise from physical laborer to successful author. It centers on the self-education, literary ambition, and romantic desire of the eponymous protagonist. After Martin saves an upper-middle-class youth from assault, he is invited to dinner and falls in love with the youth's sister Ruth. He begins a class journey that is partly fired by his desire to win Ruth and partly by his desire for knowledge and refinement. His mind is opened to science and art, grammar and etiquette, history and philosophy. The more he knows, the more powerful grows his need to express it. Though he gradually wins Ruth's love, she disapproves both of his literary ambition and of the worldview impelling it. Martin has been set ablaze by the naturalistic cosmology he first glimpses in Spencer. Through his cosmic perspective, he outgrows the upper-middle-class family who was his first contact with the cultivated world, and he becomes increasingly detached from the society they represent. His literary works are rejected repeatedly by publishers until he lives on the brink of starvation. However, his education has also separated him from his working-class companions.

When Martin's spirits are at a low, he meets a wealthy eccentric named Brissenden—a dying hedonistic poet, steeped in alcohol and drugs—who becomes his close friend. Brissenden expresses Martin's naturalistic cosmology in a poem called "Ephemera," which Martin considers the culmination of his own aesthetic ambition. In contact with Brissenden and other outcast intellectuals, Martin's naturalistic cosmology develops into an extreme individualistic outlook infused with the terminology of Spencer and Nietzsche. He attends socialist meetings for the sole purpose of attacking the speakers. This results in false reports that he is a socialist agitator, and a scandalized Ruth finally breaks off their engagement.

Shortly afterward, Brissenden dies, and the publication of "Ephemera" is met with petty incomprehension. Isolated, with both his romantic and aesthetic ambition thwarted, Martin descends into depression. Though his own literary works start to become successful at last, he has lost faith in both public opinion and society as a whole. After a brief nostalgic return to the working class and an attempt to regain happiness in Tahiti, he drowns himself.

Martin Eden is an extraordinarily ambivalent self-portrait. It is a view of London's class journey and depression filtered through his split cosmic vision. It is a fictional autobiography in which the protagonist rejects the political cause that the author believed had saved his life. Furthermore, it lacks both satirical distance and cues that the protagonist is succumbing to corruption or a tragic flaw. Underneath the self-conscious philosophical terminology and tirades against socialism, Martin simply mirrors London's own tension between individualism and prosociality. London balanced that tension by pursuing his spiritually conceived moral cause. Martin, in contrast, pursues London's egocentric naturalism to the supposedly logical conclusion of suicide.

It is little wonder that readers have been puzzled. Does Martin fail to reach for the Holy Grail, or does he defy the indifferent universe by anticipating the day of his death? Depending on which part of London's cosmic vision was in the foreground, the novel seemed in his own eyes to culminate in cosmic heroism or personal failure—in victory against the iron collar or defeat by the long sickness. To further complicate matters, most of the novel is an elated depiction of mental growth and artistic ambition, impelled by romantic love. As Sam S. Baskett observes, the novel is "a kind of portrait of the artist" (1976, 23). Martin's pursuit of love and literary expression is unequivocally aligned with London's authorial perspective. They form the emotional center of the novel and the motivational force behind the class journey it depicts. Those two pursuits span the split in London's cosmic vision: they combine a desire for naturalistic reality with a desire for moral elevation. As Martin succumbs to depression, he considers all such pursuits and desires futile. The overall meaning of his narrative depends on whether he gives up his pursuits because he succumbs to depression or whether he succumbs to depression because his pursuits are impossible. Is he a martyr to desires too great for the world, or is he not great enough to pursue his desires successfully?

The depiction of Martin certainly suggests that he is too great for the world. He is the narrative authority throughout the story, almost invariably seeing further than those around him. There are lingering descriptions of his physical attractiveness and intelligence: "bulging muscles," "curls that were a delight to any woman," and "unusual brain vigor" (1982 [1909], 574, 586, 776). His autodidact progress, modeled on London's own, is portrayed with a self-regard bordering on narcissism—and unlike London's, it is achieved with almost no guidance from friends and librarians (Haley 2010, x–7, 77–97). Martin has superhuman endurance at study and physical labor: "he never rested, and a weaker body or a less firmly poised brain would have been prostrated in a general breakdown" (1982 [1909], 643). Through his "deliberate creative genius" every story or poem is "alive in his brain, with the end in sight and the means of realizing that end in his conscious possession" (733). The greatest critics of his day are challenged by his essays, which are "brilliant, deep, philosophical, and deliciously touched with laughter" (734). Though Martin has become isolated through his intellectual development, his isolation is implicitly a result of his superiority. He concludes that no one can understand him: not his working class "gang," not his "family," not "the bourgeoisie" (888). It seems an understatement that this observation is made with a sadness "not untouched with bitterness" (888). Yet, the bitterness is not ridiculed. Martin's narrative status makes his isolation a more plausible satire of the society he rejects than of the qualities that make him reject it.

London insisted that *Martin Eden* was a critique of "Nietzschean" individualism—and, implicitly, a celebration of the socialism Martin rejects (Bruccoli 1964, vii). That claim is in line with the way he diagnosed his own depression. It has occasionally been taken at face value by scholars focusing on the class theme; the novel has even been called the archetype for stories of "unhappy upward mobility," supposedly showcasing a loss of "energy and class anger" during a tragic corruption by the bourgeoisie (Christopher 2002, 82, 87). However, the pro-socialist interpretation failed to convince London's contemporaries—including "the socialists"—for good reason (Bruccoli 1964, vii). Nietzschean philosophy and socialist theory are given hardly any detail in the novel. Like London, Martin is torn between his sympathy for the working-class and his overpowering ambition. But for Martin, the prospect of becoming a socialist warrior is peripheral to that inner tension. He never openly considers it. Brissenden urges it on him, like a prescription: "I'd like to see you a socialist ... It will give you a sanction for your existence" (1982 [1909], 853). Tellingly, this recommendation includes the same metaphor that London later used about his own depression: "you must be handcuffed to life" (1982 [1909], 853; 1982 [1913], 1066). However, Brissenden does not evoke the glory and romance of the socialist mission. He is himself a socialist only because "the slaves" are too many to allow a superman to rule: "you can't get away from them, and you'll have to swallow the whole slave morality" (1982 [1909], 853).

The confused intentions of *Martin Eden* are partly explained by its conception. London was initially writing about an ethical romance unconnected to socialist theory. The basic outline for the novel had existed in his notebooks for years as the "struggling writer idea": the story of how he began his literary career and became infatuated with the upper-middle-class girl Mabel Applegarth (Watson 1981, 400, 401–2). His first notes do not mention socialism, but they mention disillusionment and depression: the protagonist was to see "the horny side too much" and become "cynical, pessimistical" (quoted in Watson 1981, 400). When London finally wrote the novel, his peculiar interpretation of his depression had crystallized. Truth was the cause and socialism the cure. This interpretation does not shape the first part of the novel, when Martin is preoccupied with his own adventure: "he knew at last, clearly and definitely, that it was beauty, and intellect, and love that he must have" (616). When Martin is due for disillusionment, however, the only way to transform morbidity into tragedy is to give a glimpse of how he could have been saved. Martin's rise has left all other characters so far behind that Brissenden is the only one who can advise him with authority. Thus, the dubious voice for socialism becomes a nihilistic hedonist who is not only openly antisocial, but hostile to Martin's personal prosocial ideals—considering romantic love "the philosophy of madness" and literary publication "trying to prostitute beauty to the needs of magazinedom" (814, 815). The novel's intentional meaning amounts to a distinction that is both illogical in itself and awkwardly imposed toward the end of the narrative: literary publication and love are illusions that fade, socialism is an illusion that lasts.

Because of the discord between London's cosmic split and his life experience, Martin's moral trajectory is choppy and implausible. He essentially shares London's prosocial faith, imaginative elevation, and moral emotions. The core of life for him, which he aims to portray in his writing, is "moral grandeur rising out of cesspools of iniquity" (669). Like

London, he starts out seeing the cultivated world as the locus of nobility and decency; he turns Ruth into a symbol of "purity" and aspires to a "moral revolution" of physical and psychological cleanliness (579, 597). He has the "gift of sympathy," and he repeatedly extends a helping hand to his working-class fellows—most notably just before his suicide, when he donates his literary earnings to set several of them up with new careers (558). Like London, Martin struggles to translate these prosocial tendencies into rational principles. In his own assessment, he is "not a brute," and he is "a damn poor Nietzscheman"—but he does his best to be a good Nietzscheman (888). Watching the tears of his sister, he feels his reasoning "shake and totter": "the slave-class in the abstract was all very well, but it was not wholly satisfactory when it was brought home to his own family" (864). His conclusion is that he has been foolishly swayed by emotion: "pity and compassion" are "no more than the agony and sweat of the crowded miserables and weaklings" (864–65). This is a jarringly unconvincing philosophical turn. At the same time, it is neither intellectually gainsaid nor meaningfully distinct from Martin's authoritative pronouncements.

The nature of Martin's intended flaws is as hard to grasp as that of his intended virtues. His so-called Nietzschean individualism is inextricable from the exceptional qualities for which he is evidently supposed to be admired. On one of the few occasions that London interposes a distinct narrative voice, he claims that Martin was puzzled by his loneliness because "he did not know that he was himself possessed of unusual brain vigor; nor did he know that the persons who were given to probing the depths and to thinking ultimate thoughts" are to be found outside drawing rooms: "sailing solitary in the azure sky far above the earth and its swarming freight of gregarious life" (776). This supposedly detached perspective, of course, decreases the distance between author and protagonist—to the point where readers may want to look away for decency's sake. It also supports Martin's conclusion that he is too good for society. London described the socialist hero of *The Iron Heel* with the same metaphor of exceptionalism: Ernest Everhard is an "Eagle, beating with tireless wings the void, soaring toward what was ever his sun, the flaming ideal of human freedom" (1982 [1907]-a, 323). Martin's sun does not sustain him, and Brissenden is a particularly scrawny eagle. Nonetheless, in Martin's tale, Ernest's comrades come off worse.

One might wonder at London's surprise when "even the socialists missed the point" of his attack on Nietzsche (Bruccoli 1964, vii). The socialists of *Martin Eden* are such a pale presence that Martin's rejection of them is not a crisis in his development. They do not shake his Nietzschean edifice nearly so much as his sister's tears or the factory girl Lizzie Connolly's kiss. His one clash with them is memorable chiefly because of the way he attacks them. Like London, Martin sees the socialist mission as a battle against cosmic injustice. Unlike London, he sees no heroism in it:

> They were the unfit. In spite of their cunning philosophy and of their antlike proclivities for coöperation, Nature rejected them for the exceptional man. Out of the plentiful spawn of life she flung from her prolific hand she selected only the best. It was by the same method that men, aping her, bred race-horses and cucumbers. Doubtless, a creator of a Cosmos could have devised a better method; but creatures of this particular Cosmos must put up with this particular method. Of course, they could squirm as they perished, as the socialists squirmed, as the speaker on the platform and the perspiring crowd were squirming even now as they counselled together for some new device with which to minimize the penalties of living and outwit the Cosmos. (854)

Though Martin shares London's compassion for the weak throughout the novel, his vision of "the unfit" is given intellectual and rhetorical center stage in confrontation with the socialists. That vision is lingered in vividly and chillingly. The response of the socialists, in contrast, is couched in the vaguest possible terms. Martin argues in detail that "no state composed of the slave-types can endure" (855). The socialists bombard him with "lines of thought that were new to him"; they give him "new insights, not into new biological laws, but into new applications of the old laws" (856). Not a single one of these counterarguments is described. The socialists are paid homage only for their "fire and enthusiasm and excited gestures," and for their tendency to be "too earnest to be always polite" (856). They do not get to gainsay the idea that they are fighting a futile battle against the cosmos. In short, though they consider Martin "a foeman worthy of their intellect," they are not portrayed as foemen worthy of his (855).

The characterization of Martin produces an inherent queasiness that is only imperfectly recognized in London's claim to critique Nietzsche. The egocentric "Truth" of the novel is not only unhealthy and unsustainable by the human mind; it is untrue, and out of touch with the experienced reality of both author and protagonist. Neither Martin nor London lived in or wrote about a universe of isolated egos. That fact remains an undercurrent of queasiness in the novel because it is never formulated as a direct challenge. As Martin becomes ever more alienated from society, he sees that he is "very sick" (917). His sickness manifests as the fatigue, anhedonia, and motivational disruption of authentic clinical depression. He notes "apathetically" that he has stopped feeling anything: "he was not nauseated very much. He wished he could get angry, but did not have energy enough to try. He was too numb. His blood was too congealed to accelerate to the swift tidal flow of indignation" (871). He breaks his habit of four-hour rests to sleep "heavily all night"; he does not wake until roused, and remains "tired and passive"; he goes through his mail "aimlessly" and notes his publication checks "apathetically" (873). Having formerly filled every moment with goal-oriented activity, he now allows days to slip past, avoiding both people and books, "waiting for some impulse, from he knew not where"; he languishes "run down, planless, and empty and idle" (874). Life is to him a "strong, white light that hurts the tired eyes of a sick person": "It hurt. It hurt intolerably" (927). All the while, his sickness is caught up in London's account of his own depression, implying that the problem is too much Truth.

By the end of the novel, Martin takes on a fairly unequivocal function as a fictional martyr. Like London, he has no way of diagnosing his sickness except from within it: society is rotten, and all his work has been for naught. He has "hitched his wagon to a star and been landed in a pestiferous marsh"; his two greatest desires have been chimeras: the literary magazines are "all Brissenden had said they were," and he wonders "why any man wanted to marry" (875, 926). These depressive distortions insidiously merge with the authoritative cosmic vision Martin has acquired through his studies, and with his general exceptionalism. His suicide is portrayed as a majestic victory over self-preservation: he goes down "swimming with all his strength and all his will"; his last thought is a defiant restatement of the White Logic: "this hurt was not death ... it was life, the pangs of life ... it was the last blow life could deal him" (930–31). The scene recalls not only London's cool meditations on suicide as an act of freedom, but the incident (described in *John Barleycorn*) when he fell into the water as a drunken youth and suddenly saw drowning as a "fine, a splendid culminating, a perfect rounding off" of his career (1982 [1913], 993). Martin finds the romantic "key" to his suicide in "The Garden of Proserpine": "Swinburne showed him that it was the happy way out" (1982 [1909], 929). Almost since the time of

London's death, it has been debated whether he followed Martin belatedly, substituting alcohol and morphine for the Atlantic Ocean (Bruccoli 1964; Haley 2010, 313–16; Reesman 2010, 29). Indeed, London saw his alcoholism as a form of extended self-slaughter: "suicide, quick or slow, a sudden spill or a gradual oozing away through the years, is the price John Barleycorn exacts"—and it is a "just, due payment" (1982 [1913], 941).

Though Martin's morbidity is backed by all the force of London's authorial perspective, the question remains whether it is morbidity conceived as justified self-pity or as cosmic heroism. On this count the novel is genuinely ambiguous. As Baskett observes, *Martin Eden* can be taken as London's "not altogether conscious confession of a variety of personal failures, including his failure to become the iconoclastic yet affirmative writer he had expected to be" (1977, 199). There is nothing particularly heroic about the chapters that linger in Martin's depression, focusing on social isolation rather than cosmic horror: "Up above nobody had wanted Martin Eden for his own sake, and he could not go back to those of his own class who had wanted him in the past. He did not want them" (927). But Martin's death is undoubtedly framed in cosmic terms, as a clear-sighted defiance of life. In London's account of his own depression, he includes a passage that could have been a homage to Martin Eden—the cosmic hero he did not himself become:

> I still read the books, but not with the old eagerness. I still read the books to-day, but never again shall I read them with that old glory of youthful passion when I harked to the call from over and beyond that whispered me on to win the mystery at the back of life and behind the stars. (1982 [1913], 1067)

London claimed that he "no longer cared to look upon Truth naked"; he deliberately left her veils intact even when he "clutched them" in his hand (1982 [1913], 1067). But though he dwells in gratitude on the "illusion" that saved him, he is not in one mind about having pursued the lesser order of truth. In his place, Martin faces the full glory and horror of naturalistic cosmology. Martin defies life by diving into the abyss, disrobed Truth around his neck.

For London, pursuing either faith-based morality or naturalistic "Truth" was bound to be unsatisfying. Martin may partly have been a fantasy of release for the naturalism London had restrained—the way Ernest Everhard is a fantasy of a warrior's death for socialist revolution, or the factory boy of "The Apostate" is a nightmare of becoming crushed by child labor into a "travesty of the human" (1986, 219). Martin's horror vision of the socialists has a more honest ring than his stock appeals to Nietzschean slave-morality: "creatures of this particular Cosmos must put up with this particular method" (1982 [1909], 854). That phrase applies not just to the false dismissal of cooperation but to the actual cruelty and waste of natural selection. Neither Ernest nor Martin puts up with it. Ernest, like London—and like Martin's socialists—directs his ire at the method of society. Martin directs his ire at the method of the cosmos. There is an inescapably juvenile quality to the absolute rebellion of these characters; in the least generous terms of John DeCaire, they reflect London's tendency to "take his ball and go home rather than play with the cheaters" (2003, 278). The same tendency of juvenile superiority is evident in the self-praise that abounds in *Martin Eden* and *The Iron Heel*. Nevertheless, the rebellion has its roots in earnest horror arrived at through extended studies—and in an authentic experience of depression.

Before Martin is slotted into London's interpretation of his depression, he serves a more affirmative therapeutic function. His pursuit of literary publication and romantic love engages directly with London's split cosmic vision. These pursuits come together as a desire for aesthetic communion. Martin wants to share his poetic vision with both Ruth and the world. He conceives of his literary ambition as a "great," "lordly task," and it is essentially a task of communication: "to transmute feeling and sensation into speech, written or spoken, that will, in turn, in him who reads or listens, transmute itself back into the selfsame feeling and sensation" (666). Brissenden's aestheticism, in contrast, is an end in itself: "Beauty is the only master to serve. Serve her and damn the multitude!" (814). While Martin is still psychologically healthy, he urges Brissenden to "think of the rest of the world" and not to "be selfish" (833). In Martin's mind, there is no excuse for withholding works of art from the public, because "the function of beauty is joy-making" (833). His desire for aesthetic development precedes his desire for Ruth. Before he sees her, he thrills to the oil painting and books of poetry in her home like "a starving man at sight of food"; when he first sees her, she is bound up with those aesthetic experiences: "a pale gold flower upon a slender stem," "a spirit, a divinity, a goddess" who "might well be sung by that chap, Swinburne" (560). The "key" to the life he imagines with her is "reading and discussing poetry ... eternally reading and sharing" (704). But his literary ambition is first sparked by his inability to express to Ruth "the bigness and glow of life" he has seen: "it was about time for him to want to learn to talk the things that were inside of him so that she could understand" (565).

The poetic vision that Martin wants to share is essentially naturalistic. From the first, he is preoccupied with "making life as he knew it appear before his listeners' eyes" and with expressing his "love of reality" (573, 628). Before his education, he is guided by "childish notions," "half-baked sentiments," "a lot of undigested beauty," and a desire to rise in the world to marry Ruth (654). From the time he reads Spencer, he is guided by naturalistic cosmology. He starts to understand "the essential characteristics of life," pursues "the nature of beauty," and sees "a universe so concrete of realization" that it is "like the model of a ship such as sailors make and put into glass bottles" (653, 654). His first response to that naturalistic universe is not horror, but the same thrilling elevation that he feels for the art in Ruth's drawing room: he is "drunken with comprehension," "entranced by illumination"; he traces the energy of the sun through his meat and muscle until he feels "the same sun shining in his brain" (653). In the scent of a lush hill, he feels the "breath of the universe"; he sees the "romance in the life-history of any grass" and he immediately wants to translate its "hidden chemistry of sun and rain and earth" into "an epic" (665, 666). In his literary works, he aims for "an impassioned realism, shot through with human aspiration and faith," fused with "the fancies and beauties of imagination" (764). Throughout most of the novel, he wants to portray a reality that includes the moral and romantic emotions that he ends up banishing as illusions.

Martin's claim that his world crashed "with love on the pinnacle" is not very convincing (875). His frustrated idealization of Ruth is hardly a romance for the ages, or sufficient cause to lose faith in love. She falters as his ideal companion from their first meeting, when she declares Martin's beloved Swinburne "indelicate" (565). She falls decidedly short of the ideal when she cannot appreciate Martin's fiction, "attributing her incomprehension to his incoherence" (672). He realizes quite soon that his brain goes "beyond" hers, "just as it went beyond the brains of her brothers, or the brain of her father" (729). As Ruth reveals an ever-greater distance from Martin's views on life and art, he modifies his idea of love to save his love for her. He convinces himself that "love is beyond reason" and that her mind does not factor into the equation: "she was all lovable, and what she thought had nothing to do with her lovableness" (658). Yet, his romantic ideal centers on sharing and discussing aesthetic ideas:

> And always he saw that particular picture. Sometimes it was she who leaned against him while he read, one arm about her, her head upon his shoulder. Sometimes they pored together over the printed pages of beauty. (704)

After Ruth rejects him, Martin briefly considers marrying the factory girl Lizzie Connoly. He supposes that he is reluctant because of fidelity to "Love," not because Lizzie cannot "understand him nor the honor he paid her" (885, 888). But when Ruth returns repentant in the wake of his success, he rejects her as well, declaring that she would have "destroyed" his "writing" and his "career": "Realism is imperative to my nature ... all your effort was to make me afraid of life" (915). He finally realizes that he has not loved "the real bourgeois Ruth" but "an idealized Ruth": "an ethereal creature of his own creating" (916).

Martin's romantic ideal is a miniature of his general need for companionship, and the failure of his romantic ideal is a miniature of his general social isolation. The need for love that Ruth inspires in him is real enough. It is not just inspired by her poetic appearance and refined talk, but by her loving interaction with her mother: "Not in his world were such displays of affection between parents and children made":

> It was a revelation of the heights of existence that were attained in the world above. It was the finest thing yet that he had seen in this small glimpse of that world. He was moved deeply by appreciation of it, and his heart was melting with sympathetic tenderness. He had starved for love all his life. His nature craved love. It was an organic demand of his being. Yet he had gone without, and hardened himself in the process. He had not known that he needed love. Nor did he know it now. He merely saw it in operation, and thrilled to it, and thought it fine, and high, and splendid. (568)

Ruth starts out representing Martin's ethical romance and ends up representing the bourgeoisie who cannot understand him. Lizzie, in turn, represents the best of the working class. She is unafraid of life. During one of his fistfights, she watches him "with bated breath, leaning slightly forward ... one hand pressed to her breast, her cheek flushed, and in her eyes a great and amazed admiration" (883). But the only ethical romance Lizzie can manage is a pitiful offer of self-sacrifice. She could never, like Martin, have seen one of his fights as a "picture," yearning to give it aesthetic expression and "wondering if the man could paint it who had painted the pilot schooner" in Ruth's drawing room (562). Like the rest of the working class, Lizzie has become "raw" to Martin like the taste of "steam beer" (887). Between his two women—the bourgeoisie and the working class—Martin's romantic ideal never comes close to fulfillment. His interaction with the women is a narcissistic fantasy: Lizzie sobs "I could die for you. I could die for you," and Ruth implausibly promises to give up her parents and social standing: "If I have been a traitor to love, I will now, for love's sake, be a traitor to all that made that earlier treason" (888, 917). Martin's romantic ideal is greater than these interactions, but it peters out among his depressive distortions. His incompatibility with two women translates into the impossibility of romantic love.

Martin's aesthetic ambition is less ambiguously portrayed than his romantic pursuit. It is both more aligned with reality and closer to fulfillment. One might read the novel, in line with London's autobiography, as a defeat by depression. With a squint, one might read it in line with London's later dictate, as an attempt—however, conflicted and crude—to evoke the unsustainability of individualist isolation. In whatever light it is seen, however, the novel cannot be read as an indictment of the poetry, paintings, music, and prose that blaze in the protagonist's mind. Martin's responsiveness to beauty stays with him throughout the novel, and becomes a "yearning pain," a "desire to paint these visions that flashed unsummoned on the mirror of his mind" (637). As his cosmic vision expands with his education, so does his need to express himself—until, in a combination of humility and hubris, he sees himself as the voice of the universe:

> There seems to be so much in me I want to say. But it is all so big.... Sometimes it seems to me that all the world, all life, everything, had taken up residence inside of me and was clamoring for me to be the spokesman. I feel—oh, I can't describe it—I feel the bigness of it, but when I speak, I babble like a little child. (665)

Martin responds to his inner prompting by producing poems and short stories of which he is proud. He is filled with the joy of the craft, "conscious that it was great stuff he was writing" (845). He shares his work with Brissenden, who appreciates it. By the time it is embraced by the public, Martin has lost faith in public opinion. That loss of faith is an obvious distortion caused by his depression. It is even less likely, in pure numbers, that the public contains no worthy readers than that there are no worthy women for him to love. Though Martin is too dulled to notice, he has probably achieved his "lordly task."

Brissenden's poem "Ephemera" is the pivot of Martin's aesthetic ambition—and, arguably, the pivot of the narrative as a whole. For Martin, it is a more plausible Holy Grail than either Ruth or socialism. It does not fulfill his need for romantic love, but it is a culmination of the aesthetic communion underlying his romantic ideal. It is neither a personal romance detached from his mental life nor a naturalistic vision detached from personal romance, but a fully imaginative cosmic vision that affects him personally. Brissenden bases the poem on Martin's naturalistic cosmology: "your *man*, who is always the erected, the vitalized inorganic, the latest of the ephemera, the creature of temperature strutting his little space on the thermometer" (831). "Ephemera" is at once an evocation of cosmic horror and a celebration of human courage in the face of cosmic horror—a non-suicidal version of the celebration Martin represented for London. No line from the poem appears in the novel (perhaps fortunately), but Martin's response to it is evoked at length:

> It was terrific, impossible; and yet there it was, scrawled in black ink across the sheets of paper. It dealt with man and his soul-gropings in their ultimate terms, plumbing the abysses of space for the testimony of remotest suns and rainbow spectrums. It was a mad orgy of imagination, wassailing in the skull of a dying man who half sobbed under his breath and was quick with the wild flutter of fading heart-beats. The poem swung in majestic rhythm to the cool tumult of interstellar conflict, to the onset of starry hosts, to the impact of cold suns and the flaming up of nebulae in the darkened void; and through it all, unceasing and faint, like a silver shuttle, ran the frail, piping voice of man, a querulous chirp amid the screaming of planets and the crash of systems. (831–32)

The poem has a direct and physical effect on Martin. It makes his face "flushed at first," then pale; it causes his head to "swim with delight," puts "passionate tears" in his eyes, and sends "chills creeping up and down his back" (831). It is an intense version of the aesthetic and naturalistic vision of life that has been building in Martin's mind throughout the novel. For both readers and the protagonist, it stands in sharp contrast to the subsequent numbness of depression. That numbness catches hold of Martin at the very moment he registers that the public has not understood "Ephemera."

"Ephemera" is Martin's say on life as much as it is Brissenden's. After Martin has read it, he declares half in earnest that he "shall never write again" (832). More to the point, he decides to work against Brissenden's wishes so that "the great poem should see print" (844). Brissenden does not want his poem "pawed over and soiled by a lot of swine," but Martin still believes that the public can share it (833). He holds out against depression when Ruth's rejection makes him reel, and when Brissenden's death makes him wonder if he is himself "really dead and unaware of it" (869). He doggedly keeps writing, and even musters some excitement when "Ephemera" appears in print: "If only Brissenden were here to see!" (870). However, when he registers that the poem is being used for petty prestige battles—not unlike the ones for which Theodore Roosevelt had recently used London's work (Berliner 2008)—Martin breaks. He considers his own literary effort a matter of "arduous and futile years" toward inevitable failure (875). That failure severs his last emotional connection to society. He becomes "chartless and rudderless" (873–74). He begins to be haunted by a desire to escape, taking the shape of an intense vision of his sailor's days in Tahiti. Almost compulsively, he sees himself racing "on the steep face of the flying turquoise," paddling "to the pounded-coral beach where Tati's grass walls through the cocoanut-palms showed golden in the setting sun" (872). Though he claims that "the crash of magazinedom and the dear public" is insignificant next to his loss of Ruth, the turn of the narrative contradicts him (875). He tears up Brissenden's unpublished poems and does not himself write another word. Little more than a slow sinking intervenes between the end of his aesthetic ambition and the end of his life.

Martin's most constructive function for London crystallizes in his pursuit of a literary ideal. He navigates two "schools of fiction" that roughly correspond to the two halves of London's cosmic split. One that "treats of man as a god, ignoring his earthly origin" and one that "treats of man as a clod, ignoring his heaven-sent dreams and divine possibilities":

> Both the god and the clod schools erred, in Martin's estimation, and erred through too great singleness of sight and purpose. There was a compromise that approximated the truth, though it flattered not the school of god, while it challenged the brute-savageness of the school of clod. (765)

Martin's "compromise" maintains the split between the earthly and the divine—naturalism and morality—but it hints at a connection. In Martin's healthy mind, there is an approximation of truth that is less brute-savage than London's depressive Truth. It seems possible to face the naturalistic cosmos without finding life unbearable: "the more he knew, the more passionately he admired the universe, and life, and his own life in the midst of it all" (654).

Martin gives no details about his compromise between god and clod, but it is exemplified in his fiction. His stories have titles like "The Wine of Life" and "Adventure"—seemingly near-parodic London creations (667, 670). He strives for the cosmic horror and defiant heroism of "Ephemera": he wants to chronicle "the romance, and the wonder, and the glory" of creatures emerging "from out of the ruck and the mire and the countless mistakes and abortions of unending creation" (669). The stories are conceived with "fire and passion," each aiming to capture a "big thing out of life," pursuing "not the adventure of the storybooks" but "real adventure, the savage taskmaster, awful of punishment and awful of reward, faithless and whimsical, demanding terrible patience and heart-breaking days and nights of toil" (667, 670). As Martin defends his stories from Ruth's charge of "nastiness," he restates London's cosmic split, but infuses it with a meaning that approaches Stevenson's Christian cosmology:

> Saints in heaven—how could they be anything but fair and pure? No praise to them. But saints in slime—ah, that was the everlasting wonder! That was what made life worth while. To see moral grandeur rising out of cesspools of iniquity; to rise himself and first glimpse beauty, faint and far, through mud-dripping eyes; to see out of weakness, and frailty, and viciousness, and all abysmal brutishness, arising strength, and truth, and high spiritual endowment. (669)

Through this image, the horrors of the naturalistic realm are explicable as a foil for moral achievements. They are not an iron collar of mere pitiless chance, but a restriction for the sake of a system that is itself inherently good—a test for the collared souls, and a reason for the achievements of those souls to be praised. Though Martin rather conventionally places the god inside the clod—and though his faith in that arrangement is not sufficient to save his life—the passion of his rhetoric suggests that he was providing some comfort for his author. Through the originality of his framing, he may have provided some for his readers as well.

Despite the distortions in London's cosmic vision, it contains enough truth to remain both thrilling and disturbing. Wherever his imaginative influence seeps, it brings with it the grim line of Darwinian ancestry—from the "fermenting slime" that "writhed and squirmed and put out legs" through shadowy man "before he made his first tool or gibbered his first chant" (1982 [1909], 653, 772). That line carries with it the full imaginative challenge of naturalistic cosmology. And the influence seeps into unlikely places. In 1943, twenty-seven years after London's death, his life was turned into a gaudy propaganda movie emphasizing his correspondence in the Russo-Japanese war. The movie begins with the launch of the authentic WW II Liberty ship *Jack London*, framed as a symbol of American courage, and it features a scene in which a young London contemplates the brutality of his fellow sailors while reading *The Origin of Species*. The former scene (like the movie as a whole) evokes the most boyish aspect of London's adventure stories; the latter scene represents the conception of *The Sea-Wolf*, perhaps London's most direct engagement with the horror of cosmic indifference. *The Sea-Wolf* itself combines these incongruous elements: it is one of London's best-selling novels, and it is also an allegory of London's cosmic split, composed before he articulated it in his memoir or poured it bitterly into *Martin Eden*.

5.3 The Sea-Wolf (1904)

The Sea-Wolf contains a proto-Martin Eden, split into two extremes. It chronicles the maritime adventures of self-professed "sissy" Humphrey Van Weyden, a literary reviewer who is shipwrecked, picked up by a sealing schooner, and forced into servitude by the cruel Captain Wolf Larsen (London 1964 [1904], 67). Van Weyden and Wolf respectively personify London's prosocial faith and naturalistic materialism. Stripped to the bones of its thematic structure, the novel is simple and moralizing: an over-civilized man encounters the harshness of nature, survives by internalizing some of the harshness, and emerges stronger without becoming uncivilized. Wolf perishes and Van Weyden returns sea-hardened to San Francisco with a woman on his arm. London aligned himself with this stripped-down interpretation. He washed his hands of Wolf with the same words that he used of Martin: "an attack upon the super-man philosophy" (Bruccoli 1964, vii). However, as with *Martin*

Eden, London complained that "no one discovered that it was an attack upon the super-man"; in fact, *The Sea-Wolf* had earned him admiration for his "red-blooded brutality" (Bruccoli 1964, vii). It is generally agreed that Van Weyden does not dominate the novel. As Bierce declared, Wolf is "the great thing" in it, and he is a "tremendous creation" (quoted in Bruccoli 1964, vii). Nearly sixty years later, Bruccoli confirmed Bierce's assessment (and London's supposed fear): the novel "lives because it has impressed this character on the American mind" (1964, vii).

The Sea-Wolf is a fairly simple adventure-story allegory of London's cosmic vision—lacking the thematic and compositional complexity of *Martin Eden*—but an analysis of its main feature significantly illuminates how that cosmic vision was given imaginative shape. Wolf is a mythic incarnation of London's bleakest naturalistic vision. He goes much further than Martin, or even than Brissenden, as a voice of the White Logic. He is as self-serving and unapologetic as Hyde—or, indeed, as the Miltonian Lucifer to whom Van Weyden likens him. Wolf's unrestrained quality makes his Spencerian-Nietzschean egocentrism somewhat ridiculous: in Bruccoli's phrase, it comes off as "stale, bogus, or absurd." His character is in that respect on a par with the pointedly sadistic delights of Hyde. But unlike both Hyde and Lucifer, Wolf transcends the rebellion of individual assertion. His vehemence also grants power to the part of London's bleak naturalistic vision that is true:

> Why, if there is anything in supply and demand, life is the cheapest thing in the world. There is only so much water, so much earth, so much air; but the life that is demanding to be born is limitless. Nature is a spendthrift. Look at the fish and their millions of eggs. For that matter, look at you and me. In our loins are the possibilities of millions of lives. Could we but find time and opportunity and utilize the last bit and every bit of the unborn life that is in us, we could become the fathers of nations and populate continents. Life? Bah! It has no value. Of cheap things it is the cheapest. Everywhere it goes begging. Nature spills it out with a lavish hand. Where there is room for one life, she sows a thousand lives, and it's life eats life till the strongest and most piggish life is left. (1964 [1904], 49–50)

Wolf's challenge, though surrounded by facile egocentrism, is the single most powerful imaginative challenge of naturalistic cosmology: human values do not apply across the cosmos. There is no cosmic truth in the value of life, or the right to happiness, or the virtue of compassion. Such values must be found using capacities of reason and emotion that carry the potential for all human behaviors—and those capacities were themselves developed through the natural processes that involve cruelty and waste. For London, as for many of his contemporaries, Wolf's challenge was made worse by a lack of clarity about the evolutionary development of cooperation and moral emotions. But the challenge has not paled. Van Weyden, the nominal hero of *The Sea-Wolf*, is speechless before it. Like London himself, he can only declare his faith in transcendental values; like the socialists of *Martin Eden*, he asserts without specification that Wolf has read Darwin "misunderstandingly" (1964 [1904], 50). Yet, Darwin too read the cruelty and waste in nature as a book by a "Devil's chaplain" (1856).

Given the depth of despair in London's cosmic vision, it is extraordinary that he translated it into fiction that is generally accessible. *The Sea-Wolf* is significantly closer to the "adventure of storybooks" scorned by Martin Eden than it is to "Ephemera." Wolf declares war on his piratical brother Death Larsen, steals boats, quells a mutiny, and dramatically cons his cook in a card game. Van Weyden escapes with the sweet poetess Maud to play Crusoe on a Japanese island—building two separate cottages for the sake of propriety. By 1945, the novel "had sold 732,000 copies in the United States alone, many of them for 'younger readers'" (Bruccoli 1964, vi). And yet, it features an incarnation of the voice that made its author consider suicide the only freedom in life. DeCaire flippantly dismisses London's "Boys' Books of Despair"—but boys' books and despair is more apt (2003, 277). There is nothing boyish about Wolf's declaration that life is "a mess," "like a yeast, a ferment, a thing that moves and may move for a minute, an hour, a year, or a hundred years, but that in the end will cease to move" (London 1964 [1904], 37). There is nothing adventurous about his conclusion that "the lucky eat the most and move the longest"—unless it is his challenge to the ever-befuddled Van Weyden to gainsay him: "What do you make of these things?" (37). Despite inhabiting a boy's book, Wolf cuts to the quick of both cosmic meaning and the Beauty that set Martin Eden ablaze:

> "God made you well," I said.
> "Did he?" he answered. "I have often thought so myself, and wondered why."
> "Purpose—" I began.
> "Utility," he interrupted. "This body was made for use. These muscles were made to grip, and tear, and destroy living things that get between

> me and life. But have you thought of the other living things? They, too, have muscles, of one kind and another, made to grip, and tear, and destroy; and when they come between me and life, I out-grip them, out-tear them, out-destroy them. Purpose does not explain that. Utility does."
> "It is not beautiful," I protested.
> "Life isn't, you mean," he smiled. "Yet you say I was made well." (101)

Wolf's biology reaches all the way to aesthetics. Martin's cosmic thrills—the very "color and flashing light" that he saw through Swinburne's eyes—are implicated in the process of natural selection. Though Wolf emphasizes the cruelty of nature and leaps illogically from naturalism to egocentrism, his blunt-force metaphysics are remarkably accurate. From the perspective of human morality, natural selection is revolting; yet, natural selection has shaped the human perspective and its perception of the beautiful or good. Life is not beautiful; yet you say I was made well.

Wolf's ambiguous narrative status has occupied scholars as much as that of Martin Eden. In 2010, Eric Carl Link noted that *The Sea-Wolf* "continues to be the object of considerable critical speculation, with some several dozen articles and book chapters" devoted to it "during the past few decades" (2010, 150). For Link, the novel's ambiguity centers on "the curious nature of Wolf Larsen's death" (2010, 150; 2013, 171). Wolf's death is indeed curious, and the debate surrounding it is a neat miniature of the debates about his overall narrative meaning. Half a decade before Link, Bruccoli's authoritative introduction to the novel zeroed in on the same problem: in symbolic terms, he claims, "there is no explaining Larsen's fate" (1964, xv). According to Bruccoli, Wolf's death is artistically incoherent because he is not killed by his brother Death, or by Van Weyden, or by the sea, but by a brain tumor that deprives him "of the physical and mental qualities with which he had successfully defied nature" (Bruccoli 1964, xv). Link explains Wolf's death as an illustration of Haeckel's dismissal of egocentrism, and provides five possible metaphorical meanings: the self-destruction of a Nietzschean superman, the inflexible brute losing a Darwinian struggle, the evil man struck down by God, the gothic figure hunted by his doppelganger (Death Larsen), and the anachronism destroyed by capitalist competition. These interpretations are based on a survey of interpretive readings of the novel (Link 2010, 153–58). They all stress Wolf's iconoclastic individualism, and they hint at real themes in the novel. But none of them makes sense of Wolf's materialistic cosmic challenge—the unanswered challenge that makes him more than a stock villain, and that holds the novel's protagonists suspended outside civilization until they drop his corpse into the sea (spotting their rescue on the horizon as they turn from the task).

Framing Wolf as arch-villain, allegory of the unfit, or victim of social systems does not explain why he remains "sceptical and invincible to the end" when his body shuts down sense by sense, or why he scribbles an answer to Maud's final insistence on immortality to which London would have raised a glass: "B-O-S–H" (248). Those features make sense in the light of London's imaginative response to Darwinism: his split cosmic vision and his oscillation between heroic Truth and healthy illusion. From that perspective, Wolf's fate is both explicable in its ambiguity and, within its range, symbolically ingenious. Wolf pursues Truth single-mindedly and is destroyed by his own constitution. Like Martin, he is eaten up and ultimately killed by a paralyzing growth in his brain. His first blinding headache is directly linked to his "blue moods," his "primal melancholy" and his "sadness as deep-reaching as the roots of the race" (68). Per Serritslev Petersen comes close to this explanation when he claims that Wolf's brain tumor appears "psychosomatically," but he attributes it to the influence of Nietzsche without explaining the psychological appeal or effect of the ideas themselves—or, indeed, acknowledging their origin (for London and the world) prior to Nietzsche (2011, 69).

Within *The Sea-Wolf*, both Wolf and Van Weyden spell out Wolf's narrative meaning in terms of London's cosmic vision. Van Weyden diagnoses the case correctly despite himself being safely sense-struck: "The old primal melancholy was strong upon [Wolf] ... He had reasoned himself into a spell of the blues" (158). Van Weyden sadly concludes that Wolf's "brutal materialism" will not allow him "the chief vent to this primal melancholy," which is "religion" (68). Like Martin, Wolf is not purely depressive. Though he has reasoned away the lasting meaning of constructive creativity, social bonds, and kinship ties, he feels the flickering joy of life—the lure of the healthily sense-struck "lesser order of truth." He compulsively pierces it with his mind:

> "Do you know, I am filled with a strange uplift; I feel as if all time were echoing through me, as though all powers were mine. I know truth, divine good from evil, right from wrong. My vision is clear and far. I could almost believe in God. But," and his voice changed and the light went out of his face,—"what is this condition in which I find myself? this joy of living? this exultation of life? this inspiration, I may well call it? It is what comes

> when there is nothing wrong with one's digestion, when his stomach is in trim and his appetite has an edge, and all goes well. It is the bribe for living, the champagne of the blood, the effervescence of the ferment—that makes some men think holy thoughts, and other men to see God or to create him when they cannot see him. That is all, the drunkenness of life, the stirring and crawling of the yeast, the babbling of the life that is insane with consciousness that it is alive. And—bah! To-morrow I shall pay for it as the drunkard pays. And I shall know that I must die, at sea most likely, cease crawling of myself to be all a-crawl with the corruption of the sea; to be fed upon, to be carrion, to yield up all the strength and movement of my muscles that it may become strength and movement in fin and scale and the guts of fishes. Bah!" (54)

It is difficult to imagine the adventure-greedy boy devouring this "swift rush from sublime exultation to despair" (54). Literary scholars have in the main refused it too, construing Wolf's cosmic horror as incidental to themes of individualism and prosociality, with or without political connotations (Link 2010; Mills 1995; Pizer 1983). In that social interpretation, Wolf is simply expressing a mistaken view as part of the "symbolic condemnation of egoism left unchecked by altruism" (Link 2010, 163). Wolf, in contrast, sees his situation as London conceived it: "He who delights the most lives the most, and your dreams and unrealities are less disturbing to you and more gratifying than are my facts to me" (157).

Wolf is the knife of cosmic horror pressed against London's throat. Through his repeated requests for answers, he is also London's imagination "cursed" to discover the cosmos. He is plagued by "curiosity," by a "questing," an "everlasting query" as to "what it was all about" (62). He is both the mystery of life and the tortured search for answers. His eyes are of a "baffling protean gray," like the shifting fog which in the beginning suggests to Van Weyden the "shadow of infinite mystery,"

> brooding over the whirling speck of earth; and men, mere motes of light and sparkle, cursed with an insane relish for work, riding their steeds of wood and steel through the heart of the mystery, groping their way blindly through the Unseen, and clamouring and clanging in confident speech the while their hearts are heavy with incertitude and fear. (5)

Wolf embodies the undercurrent of tortured human consciousness that people like Van Weyden can only glimpse. His unapologetic brutality makes him incline more toward cosmic law than consciousness: "He's not black-hearted like some men. 'Tis no heart he has at all. Wolf, just wolf, 'tis what he is" (42). But because he is a human incarnation of these natural forces, he is held accountable for their cruelty. He is nature as the "enslaver and tormentor of men": "a male Circe and these his swine, suffering brutes" (170–71). On the cosmic level, he half becomes an evil god of naturalism, therapeutically to be hated and destroyed. On the human level, Van Weyden dismisses him as "a man utterly without what the world calls morals," "a sort of monster," "living an obscure and sordid existence ... that is anything and everything except splendid" (59, 60, 71). Wolf is declawed, physically paralyzed, and buried deep in the sea: "Good-by, Lucifer" Maud whispers—but she adds "proud spirit" (255). And in the minds of readers, Wolf is not dead.

London, unlike Wolf, was not a "human brute" (23). This is a difference in kind between author and character: Compassion is "like sea-sickness, I suppose," Wolf observes, "some men are subject to it, and others are not" (49). During London's days as a tramp, he was disturbed to the core by watching a child and mother being abused: "in his eyes was a terror that made me sick—the frantic terror of an inconceivably maltreated child"; "the sweat poured from me, and I breathed hard, clutching the grass with my hands until I strained it out by the roots" (1982 [1907]-b, 223–24, 225). The incident is evoked at length, and he compares it to the time he witnessed a hanging, when his "whole soul cried protest" (1982 [1907]-b, 224). In short, like Van Weyden, and like Martin Eden, he was made "queasy" by brutality (1964 [1904], 49). But even with this pointed distinction, *The Sea-Wolf* aligns with the Devil's party more readily than *Paradise Lost*. It is a simpler and more intense expression of the horror and heroism in *Martin Eden*. London himself liked to be called Wolf (Haley 2010). Three years before he wrote *The Sea-Wolf*, he asked a correspondent "after all what squirming anywhere, damned or otherwise, means anything?"—"That's the question I am always prone to put: What's this chemical ferment called life all about?" (quoted in Petersen 2011, 68). Toward the end of his life, while he persisted in washing his hands of his "super-men," he built a Wolf House on his Beauty Ranch. For symbolic consistency—or perhaps to repay his Pilate act—the house burned to the ground before it could be occupied.

5.4 Conclusion

The "Adventure" that Jack London lived and wrote frames Darwinism with an astonishing combination of fortitude and fantasy. His celebrations of meritocratic success—whether of species, ethnicities, or individuals—contain nothing of Doyle's certitude. He could not cling to biology as a means of utopian improvements, and, despite Martin's saints in slime, he could not rest like Stevenson in the metaphysical dualism of Christianity. He internalized the cosmic horror of naturalism, but he distorted it into a vision of egocentric isolation, despaired, and disavowed it. When he saw suffering, he told reality to go hang for not conforming to the Golden Rule, but he equally accused his readers—"O dear, soft people, full of meat and blood"—of not facing the reality of suffering (1982 [1903]-a, 46). He declared socialist revolution a panacea and an impossibility and an inevitability. His artistic ambition scorched him—like "an everlasting pain," Brissenden says, "a wound that does not heal, a knife of flame"—but he ended up mass-producing adventure literature and discounting the role of art in human life: biologically, as an absurdity; politically, as a tool of power (London 1982 [1909], 814; 1986, 298–312).

London's lumpy imaginative concoction has been an irritant over the years. Since the mid-twentieth-century, the critical conception of his fiction has slipped from esteemed artistry to boy's adventure. James Haley argues that the Red Scares of the 1920s and 1950s unfairly branded London as a writer of "juvenile adventure"—supposedly as a declawing measure because he was too popular to erase (2010, x). This explanation may not be wholly outlandish. London's pursuit of his revolutionary Grail called his American loyalties "so much into question that, though he was long dead, the FBI opened a dossier on him" (Haley 2010, x). However, disparagements of his literary qualities come from across the political spectrum. In 1941, George Orwell paid a grudging tribute to the prediction of fascism in *The Iron Heel* (1908)—while declaring it "crude" and implying that London himself had "a fascist streak" (Orwell 1941). In 1964, Matthew Bruccoli—though calling London a "serious writer" and sympathizing with his socialism—attributed his declining critical acclaim to the "dated" quality of his work (Bruccoli 1964, vi). At the same time, London kept finding readers. *Call of the Wild* and *White Fang* remain household names. Many of London's short stories are beloved as adventure literature—and some of the same stories, like "To Build a Fire," "The

Chinago," "A Piece of Steak," "Love of Life," and "The White Silence," display the stylistic and thematic complexity of high literature.

In 1983, Donald Pizer formulated the literary critic's dilemma directly: "to explain how a writer who could achieve the seamless perfection of 'To Build a Fire' could also produce an extraordinary amount of trash" (107). Recent scholars have made it their business to vindicate London as a "Great Writer" (Haley 2010, xi–xii). They have argued plausibly that he produced "works of stunning originality and descriptive power"—works that stung "the nation's conscience for its callous disregard of its most vulnerable members" or that displayed "his deeply studied craft of the novel wedded to his own high adventures"—and at the same time, they have had to admit that he produced self-confessed "hackwork," riddled with absurd plots and melodramatic characters (Haley 2010, xii). London variously wanted his fiction to inspire political change, to make him a quick buck, or to conjure the cosmic aesthetic chills of "Ephemera."

Through his stratospheric ambition and the hard work that sustained it, London achieved a facility that made him extraordinarily prolific. The prolific output meant a natural percentage of infelicity, but the facility left his imagination free to infuse occasional adventure tales with cosmic significance: to "utter the universal thing" beneath "the swing and go of the story" (1982 [1909], 844). In Emma Darwin's phrase, his work (or some of it) "can be read"; and when it is, the best of it leaves an impression more mythic than boyish. When he expressed his cosmic vision in direct allegory, as in *The Sea-Wolf*, he was hampered by the falsehoods of his cosmic split; when he inserted himself as the hero, as in *Martin Eden* and *The Iron Heel*, he was hampered by an understandably but unattractively high self-regard. "To Build a Fire" has an almost rudimentary plot, chronicling a man fighting the cold on the Yukon trail. Beneath its swing and go of hope and horror, it is a vision of London's naturalistic humanity—mortal, endlessly vulnerable and endlessly venturing, separated from the certainty of instincts to make its way through a hostile world. It has the beauty of Martin's aesthetic thrills and the horror of Wolf's sneer without the ruptured seams of either.

London's focus on survival in extreme circumstances may have helped him to become generally accessible. That is a theme fully consistent with the adventure story genre, and it may allow readers to conceive of the Darwinian struggle as a special case outside the bounds of human civilization. His declarations of prosocial faith were no doubt taken at face value by those of his readers who never glimpsed "the truer order of truth"

(1982 [1913], 1094). More importantly, London's joy in life often overcame his despair, and one of the things he found joy in was the reality of human sociality. He channeled the imaginative threat of Darwinism, but also the excitement of its explanatory power; he reveled in its suggestions of strength and beauty arising out of slime, and of truths deeper than conventional mores. His fiction often displays a cosmic horror taking the shape of merciless suffering or ruthless competition, but it also displays the pleasures of camaraderie, the pathos of compassion, and a sense of justice that variously appears as boyish idealism or socialist doctrine. At their best, his stories both depict and embody the triumph of worthy achievement under circumstances excruciatingly harsh.

More often than not, London's heroes mirror Martin Eden's saints in slime. They respond to extreme circumstances with compassion or high principles. "Love of Life," one of his grimmest short stories, reaches a pitch of implausible poetic justice when the starving protagonist comes across the remains of his partner: "Well, Bill had deserted him; but he would not take the gold, nor would he suck Bill's bones. Bill would have, though, had it been the other way around" (1986, 137–38). During a life-or-death trek in "The White Silence," the Klondike hero Malamute Kid—"strong man, brute that he was, capable of felling an ox at a blow"—refuses to gain speed by whipping his dogs: he "could not bear to beat the poor animals, but humored them as a dog-driver rarely does,—nay, almost wept with them in their misery" (1986, 14). The protagonist of "To Build a Fire," freezing to death after ignoring the advice of a local elder, ends his life on a quiet concession: "You were right, old hoss; you were right" (1986, 178). In "The Red One," the naturalist Bassett, dying of fever among headhunters who worship an extraterrestrial device, decides not to ruin his ritual decapitation by shooting himself in the head: the head-curer has "played squarer than square," behaved as "a forerunner of ethics and contract," and so "it would be a ghastly pity and an act of dishonor to cheat the old fellow at the last" (385). These characters consistently find themselves in circumstances that bear out Martin Eden's suggestion that "life is not always beautiful" (1982 [1909], 668). Though London found no stability in cosmic values or the axioms of faith, he repeatedly found footing in small acts of selflessness or integrity performed under duress—acts that did not uphold the weight of cosmic virtue and sin, but displayed a type of human behavior that was more morally evocative than others. To find his models, he consistently looked where life was not beautiful: "Perhaps because I am strangely made, I find something beautiful there. It seems to me that the beauty is tenfold enhanced because it is there" (1982 [1909], 668).

London internalized the Darwinian struggle and visualized it on a cosmic scale. Through the expanse of deep space, he traced deep time and "biology in its largest aspects": "the very stuff out of which has been spun the fabric of all the arts, the warp and the woof of all human actions and achievements" (1982 [1909], 772, 773). Like Brissenden, he heard "the screaming of planets and the crash of systems," interspersed with "the questing, eternal, ever recurring, thin little wailing voice of man" (1982 [1909], 831, 832). Sometimes, like Wolf, he approached this wailing creature with "the cruel hand of a vivisectionist" (1964 [1904], 54). Sometimes, like Martin, he felt himself a piece of "the same nonunderstandable fabric, twisted of sunshine and star-dust and wonder" (1982 [1909], 734). Like Van Weyden, he clung to principles of decency without being able to articulate or rationalize them, and he came to rest in a smile "whimsical with love" (1964 [1904], 256). Like Basset, he professed scientific detachment but appealed to the cosmos for answers and guidance: "these far distant and superior ones ... Had they won Brotherhood? Or had they learned that the law of love imposed the penalty of weakness and decay? ... Was the rule of all the universe the pitiless rule of natural selection?" (1986, 381). These characters bring alive the intellectual and imaginative struggles of a time still raw with the wound of naturalistic cosmology. The rawness, more or less rudely traced, provides the outline of our present world-dramas. The mental processes of London's saints and narcissists provide clues to our own scrambles for personal motivation and social values in a naturalistic world. "We enter into the minds of the actors in that long-ago world-drama, and for the time being their mental processes are our mental processes" (1982 [1903]-b, 320).

References

Baskett, Sam S. 1976. "'Martin Eden': Jack London's Poem of the Mind." *Modern Fiction Studies* 22 (1) 23–36.

———. 1977. "*Martin Eden:* Jack London's 'Splendid Dream'." *Western American Literature* 12 (3): 199–214. https://doi.org/10.1353/wal.1977.0013.

Berkove, Lawrence. 2004. "Jack London and Evolution: From Spencer to Huxley." *American Literary Realism* 36 (3): 243–55.

Berliner, Jonathan. 2008. "Jack London's Socialistic Social Darwinism." *American Literary Realism* 41 (1): 52–78. https://doi.org/10.1353/alr.0.0003.
Bruccoli, Matthew J. 1964. "Introduction." In *The Sea-Wolf* by Jack London, edited by Matthew J. Bruccoli, v–xv. Boston: Houghton Mifflin Company.
Christopher, Renny. 2002. "Unhappy Narratives of Upward Mobility: *Martin Eden, Bread Givers, Delia's Song*, and *Hunger of Memory*." *College Literature* 29 (4): 79–108.
Conrad, Joseph. 2015. *The Selected Letters of Joseph Conrad*, edited by Laurence Davies. Cambridge; New York: Cambridge University Press.
Darwin, Charles. 1856. "Letter to J. D. Hooker on 13 July 1856." *Darwin Correspondence Database*. "Letter no. 1927." University of Cambridge. https://www.darwinproject.ac.uk/letter/DCP-LETT-1924.xml.
DeCaire, John. 2003. "The Boys' Books of Despair." *Southwest Review* 88 (2/3): 227–90.
Faulstick, Dustin. 2015. "'The Preacher Thought as I Think': Wolf Larsen, Humphrey Van Weyden, and Jack London's Ecclesiastes." *Studies in American Naturalism* 10 (1): 1–21. https://doi.org/10.1353/san.2015.0004.
Haley, James L. 2010. *Wolf: The Lives of Jack London*. New York: Basic Books.
Harpham, Geoffrey Galt. 1975. "Jack London and the Tradition of Superman Socialism." *American Studies* 16 (1): 23–33.
Link, Eric Carl. 2010. "The Five Deaths of Wolf Larsen." *Studies in American Naturalism* 5 (2): 151–63. https://doi.org/10.1353/san.2011.0006.
———. 2013. "Jack London, *The Sea-Wolf*, and the Natural History of Love." *Studies in American Naturalism* 8 (2): 171–85. https://doi.org/10.1353/san.2013.0018.
London, Jack. 1982 [1903]-a. "The People of the Abyss." In *Novels and Social Writings*, edited by Donald Pizer, 1–184. New York: The Library of America.
———. 1982 [1903]-b. "How I Became a Socialist." In *Novels and Social Writings*, edited by Donald Pizer, 1117–20. New York: The Library of America.
———. 1964 [1904]. *The Sea-Wolf*, edited by Matthew J. Bruccoli. Boston: Houghton Mifflin Company.
———. 1999 [1906]. "What Life Means to Me." In *No Mentor but Myself: Jack London on Writers and Writing*, edited by Dale L. Walker and Jeanne C. Reesman, 87–94. Stanford, CA: Stanford University Press.
———. 2016 [1907]. *Before Adam*. London: CreateSpace Independent Publishing Platform.
———. 1982 [1907]-a. "The Iron Heel." In *Novels and Social Writings*, edited by Donald Pizer, 315–553. New York: The Library of America.
———. 1982 [1907]-b. "The Road." In *Novels and Social Writings*, edited by Donald Pizer, 185–314. New York: The Library of America.

———. 1983 [1908]. "A Curious Fragment." In *Science Fiction: A Historical Anthology*, edited by Eric S. Rabkin, 208–18. New York: Oxford University Press, 1983.

———. 1982 [1909]. "Martin Eden." In *Novels and Social Writings*, edited by Donald Pizer, 555–931. New York: The Library of America.

———. 1982 [1913]. "John Barleycorn." In *Novels and Social Writings*, edited by Donald Pizer, 933–1112. New York: The Library of America.

———. 1986. *To Build a Fire and Other Stories*. New York: Bantam.

London, Jack, and Anna Strunsky. 1903. *The Kempton-Wace Letters*. New York and London: Macmillan.

London, Joan. 1939. *Jack London and His Times: An Unconventional Biography*. New York: University of Washington Press.

Mills, Gordon. 1995. "Jack London's Quest for Salvation." *American Quarterly* 7 (1): 3–14. https://doi.org/10.2307/2710410.

Orwell, George. 1941. "Wells, Hitler and the World State." *Horizon* 4 (20): 133–38.

Petersen, Per Serritslev. 2011. "Jack London's Dialectical Philosophy Between Nietzsche's Radical Nihilism and Jules de Gaultier's *Bovarysme*." *Partial Answers: Journal of Literature and the History of Ideas* 9 (1): 65–67. https://doi.org/10.1353/pan.2011.0009.

Pizer, Donald. 1983. "Jack London: The Problem of Form." *Studies in the Literary Imagination* 16 (2): 107–15.

Reesman, Jeanne Campbell. 2010. "The Call of Jack London: Earle Labor on Jack London Studies." *Studies in American Naturalism* 5 (1): 21–36.

Watson, Charles N. 1981. "The Composition of *Martin Eden*." *American Literature* 53 (3): 397–408. https://doi.org/10.2307/2926227.

CHAPTER 6

H. G. Wells's Evolutionary Imagination

> There is ... a sense of infinite peace and protection in the glittering hosts of heaven. There it must be, I think, in the vast and eternal laws of matter, and not in the daily cares and sins and troubles of men, that whatever is more than animal within us must find its solace and its hope. I hope, or I could not live.
>
> —H. G. Wells, *The Island of Dr Moreau*

> My reply to the superior critic has always been—forgive me—"Damn you, do it better."
>
> —H. G. Wells, *World Brain*

If the late-Victorian literati were allowed only one "evolutionary myth-maker," the title would go to H. G. Wells. Like Jack London, Wells has a peculiar place in literary history. He had ambitions as a scientist, a realist novelist, a satirist, a journalist, a global political reformer, and a historian. His greatest ambitions were toward political reform, but history has judged otherwise. He is now remembered almost exclusively for the early fantastical stories he dubbed *scientific romances*. In those stories, he used his education in natural science to invent arresting fictional scenarios, to imagine futures full of previously unimagined technology, and to paint humanity's Neanderthal-haunted past. Inflamed by T. H. Huxley's biology classes, he conjured unseen life—often sentient—from alternate

E. Jonsson, *The Early Evolutionary Imagination*,
Cognitive Studies in Literature and Performance,
https://doi.org/10.1007/978-3-030-82738-0_6

dimensions, other planets, and the unexplored depths of the sea. This imaginative explosion helped transform the Victorian adventure genres into scientific mythology. Wells has been called a "'founding father' of science fiction" and the "Shakespeare of science fiction" (McLean 2011, 476; Aldiss 1986, 166). Novels like *The Time Machine* (1895), *The Island of Doctor Moreau* (1896), *The War of the Worlds* (1897), and *The First Men in the Moon* (1901) are emulated and alluded to by subsequent scientific romancers, "all of whom"—as Brian Aldiss observes from experience—"stand in Wells's portly shadow" (Aldiss 1986, 152). But Wells himself considered these novels "playful parables": relatively unimportant outcroppings of his mission to guide the world toward a modern utopia (1978 [1934], vi). There was a time when his judgment seemed right.

For intellectuals during the first half of the twentieth century, Wells's ideas about the world were inescapable. He spoke on radio and made films, produced biology textbooks and drafted Human Rights declarations, advocated for the League of Nations (then turned against it), and traveled between the White House and the Kremlin to personally connect the global powers (Aldiss 1986; Wells 1934; Orwell 1941). All the while, he churned out best-selling non-fiction: *Mankind in the Making* (1903), *New Worlds for Old* (1908), *War and the Future* (1917), *The Outline of History* (1920), *The Salvaging of Civilization* (1921), and *The Work, Wealth and Happiness of Mankind* (1931). These brashly titled books were all premised on his ideal of a "planned world," or World State.

Though the World State mutated over the course of Wells's life, its core remained the same: a worldwide organization, centered on education and exploration, run by a scientific council (Wells 1902, 1934, 549–703). It was a recognizable constant for his contemporaries. In 1909, it provoked E. M. Forster into the uncharacteristic science fiction satire "The Machine Stops," in which a global technocratic state crumbles beneath the "untainted sky" (Forster 2011 [1909], 56). Two decades later, Aldous Huxley's dystopian Dr. Wells administered pseudo pregnancies to the ladies of *Brave New World* (2007 [1932], 32). In 1941, George Orwell complained that Wells was trying to banish Hitler with "the usual rigmarole about the World State":

> Except that he is now especially concerned with federal world control of air power, it is the same gospel as he has been preaching almost without interruption for the past forty years, always with an air of angry surprise at the human beings who can fail to grasp anything so obvious.

Orwell does not object to Wells's ideal in itself. It is a "common-sense, essentially hedonistic world-view": "All sensible men for decades past have been substantially in agreement with what Mr. Wells says" (1941). So why had Wells's social vision not been realized? Orwell, who would produce *1984* seven years later, identifies the problem with unsurprising acuity: "the sensible men have no power, and, in too many cases, no disposition to sacrifice themselves... The energy that actually shapes the world springs from emotions—racial pride, leader-worship, religious belief, love of war—which liberal intellectuals mechanically write off as anachronisms" (1941).

The World State manifesto wrote these emotions off by declaring that they would have to be suppressed through education (1975 [1897]; 1934, 558, 559). Wells tended to discount the potential positive effects of cooperative or imaginative emotions. And he did so for the same reason that Jack London blinded himself to moral passions: he drew a sharp line between the natural and cultural parts of human experience. In essence, Wells shared his teacher T. H. Huxley's idea that nature is egocentric and that ethics is a non-natural cultural force (Huxley 2009 [1893]; Wells 1975 [1897]). The World State, as an extreme of Huxley's ideal, was a new form of religion for Wells: a supernatural society supported by a supernatural force of moral education (Wells 1902, 1915). He believed that humanity should use moral education to correct religion, tribalism, and patriotism. Unlike London, he had little patience for those who wanted to remain uncorrected, or "life-blinded and sense-struck" (London 1982 [1913], 1094). In fact, Wells believed that such people would cause the destruction of humanity. Through the shifts and inconsistencies of his outlook, he remained dedicated to the ideal of scientific education—and he remained convinced that it was the only thing that could save our species from self-destruction (Di Leo 2017; Earle 1950; Wagar 1989). He sums up his diagnosis in *The Outline of History:* "Human history becomes more and more a race between education and catastrophe" (1921, 1100).

Wells's fiction obsessively evokes his idea that our species will perish without education and social planning. Though he became increasingly propagandistically plain, the message was present in his earliest scientific romances. As he put it, he had simply begun by discussing "sociology in fable" and addressing the "problem of human will and government under fantastic forms" (1934, 558, 559). The fantastic forms were particularly appropriate for his evolutionary nightmares. One of his most striking depictions of human catastrophe appears in his first novel. In *The Time Machine* (1895), "the human intellect" has "committed suicide" by achieving perfect comfort and short-term happiness—causing our species to diverge into ape-like predators and their doll-like cattle (2005 [1895], 57). Wells remained preoccupied with this nightmare until, in his final book, he gave up: he decided that the human species would either perish or become unrecognizable (1946).

As long as Wells believed that the human catastrophe was preventable, he wrote with an urgency that was occasionally callous. Forty years after *The Time Machine*, in the movie script for *Things to Come* (1936), he put the case plainly to his audience. In a future where the World State has been achieved, humanity still needs to save itself from evolutionary degeneration by meeting the continuous challenge presented by the scientific exploration of space. The leader of the World State sets an example by overcoming personal emotion, risking his own child and the child of his friend Passworthy on the first moon voyage. He ends by imploring humanity not to be "little animals" who live for short-term happiness:

> If we are no more than animals, we must snatch at our little scraps of happiness and live and suffer and pass, mattering no more than all the other animals do or have done. [He points at the stars.] It is that—or this. All the universe—or nothingness ... Which shall it be, Passworthy?

Wells's script ends with the question reverberating among the stars, its disembodied voice rising on the darkening screen: "Which shall it be?" It was a question he posed to most of his audiences in some form. And he often posed it harshly. From his cosmic vantage point, he could declare with perfect sangfroid that the World State would eliminate those humans—individuals and populations—who failed to live up to the "new needs of efficiency" (1902, 317).

To literary scholars, Wells poses several related challenges: his semi-canonical status as a science fiction writer, his broad scope and implausibly prolific output, his didactics based on eclectic politics, and his harsh Darwinian social vision. These features make him hard to pin to any particular ideology or cultural school. The task is not simplified by Wells's tendency to use "Socratic dialogue and similar rhetorical devices," to state "a point of view more effectively than its own proponents" before "demolishing it" (Earle 1950, 184). During his vocal career, he often could not be found in the last position he had defended, because—in

Duncan Bell's understated phrase—"He had moved on" (2018a, 422). Scholars trying to capture Wells's general shape resort to a distinct species of words: "protean" (Aldiss 1986, 153; Bell 2018a, 1), "amorphousness" (Aldiss 1986, 153), "conflicted" (Collinge 2017, 47), "confused," "mixed agendas" (Glendening 2002, 571), "polymath writer," "diverse interests" (Beenstock 2015, 57), "paradoxical and volatile," "versatile and mercurial" (Earle 1950, 183, 185). Nevertheless, scholars have converged on a basic image of a politically engaged, ambitious scientific rationalist, with a gift for imaginative speculation, riding an undercurrent of intellectually unacknowledged emotion (Bell 2018a, b; Bloom 2005; Huntington 2005; Wagar 1989; McLean, 2011). Wells's political worldview has been described in great detail since the middle of the twentieth century (Bell 2018a, b; Earle 1950; Wagar 1989), and his obsession with evolutionary theory has been widely acknowledged (Barnett 2006; Glendening 2002; Hillegas 1967; Wagar 1989).

The challenge in explaining Wellsian fiction is not just to see its prominent parts. It is to see all of those parts at once, in proportion to each other, and to fashion a coherent explanation of their psychological functions and effects—to produce, so to speak, an Outline of Wells with a scientific foundation. Rather than pursuing that project, literary scholars have seemed to follow the protagonist of Wells's satirical bildungsroman *The Bulpington of Blup* (1932). The eponymous Bulpington, faced with the world riddle, resorts to isolated impressions: "What was the Real Truth of Things? It seemed to be very difficult to get to it while one looked at everything at once" (38). Many scholars faced with the Wells riddle have focused on single cultural influences: empiricist political theory, American pragmatist philosophy, the zoologist Edwin Ray Lankester, Liberal eugenics, 1890s debates about primitivism, Plato, or the Victorian suffrage movement (Beenstock 2015; Bell 2018b; Barnett 2006; Maxwell 2012; McNabb 2015; Thompson 2017; Pudelko 2015). Others have gone to the same extreme as the impressionistic Bulpington, and looked for Truth by turning "inward" to their own preoccupations—using Wells's novels as cues for contemporary reflections on class, race, and gender (Wells 2008 [1932], 38; Bloom 2005; Collinge 2017; Higgins 2008; Hume 1990; McLean 2011).

Though large and complex subjects must be analyzed in parts, isolating the parts comes with a risk of expanding each one disproportionately. This risk is evident in scholarship on Jack London, whose sprawling life and inner conflicts transcend readings of him as a "socialist," "Nietzschean," or "socialist Nietzschean." When dealing with Wells—who spent more years on the international cultural stage than London spent alive—the risk is greater still. Scholars who use his novels as cues for ideological reflections embrace this risk. They often end up beyond the realm of plausibility, as when Richard Higgins argues that Wells depicts violent emotion because of his class journey, or James T. Collinge explains *The Time Machine* and *Doctor Moreau* through Wells's supposed attitude to rabbit poaching (Higgins 2008; Collinge 2017). Such scholars rarely find common ground with each other. Without seeking integration with the interpretations based on class, Kathryn Hume reads *The Time Machine* as a novel about Freudian oral fantasies culminating in the indomitable feminine, and John Glendening argues that *Moreau* allegorizes Derridean confusion as seen through Darwin's entangled bank (Hume 1990; Glendening 2002). Whatever insights emerge from such readings—the truth of Wells's defensive class consciousness, his absorption in local political issues, his unacknowledged erotic turmoil, or his conflicted emotions about ecological interaction—those insights would appear more clearly if they were presented in proportion to the concerns that dominated Wells's imagination.

The grand vision of the World State is Wells's story about himself. As noted in 1950 by Edward Mead Earle—American political scientist and military strategist—it is the key without which "no one can hope to understand Wells" (183). Wells himself describes it as the crystallization of his intellectual life and the foundation for any claim he has on the public's attention: "Either the toad which is struggling to express itself here, *has* engendered a jewel in its head or it is nothing worth troubling about in the way of toads" (1934, 11). Beneath his global vision, germinating during the intellectually formative years that produced his enduring science fiction, lies the dual view of humanity that was formed in response to the psychological challenges of Darwinism. "Natural man" and "artificial man"—egocentric nature and restraining culture—are the mythic creatures on whose backs Wells rests his imaginatively tolerable natural world (Wells 1975 [1897], (1975 [1897]). Those mythic creatures fragment into their most memorable shapes in his early fantastical novels. There, still unrestrained by the full articulations of the World State, they interbreed with Wells's intuitive understanding of moral and imaginative emotions. They mutate into nightmares that sustain his utopian dreams: the various failures of humanity, hyper-domesticated, hyper-rational, or uneducably bestial.

Though the universal challenges of Darwinism illuminate Wells's fiction, no one has faced Darwin in quite the same way as he did. Wells's personality shaped his fiction so powerfully that Henry James, his some-time friend, declared that "the ground of the drama is somehow most of all in the adventure for *you*—not to say *of* you, the moral temperamental, personal, expressional, of your setting it forth" (Wells 1934, 413). The features of his personality can be gleaned from biographies, contemporary accounts, and his own substantial reflections on the subject: a contrarian but gregarious social temperament, a certain shameless calm, an intense ambition threatened by hedonistic dissolution, a deep need for truth and cosmic meaning, an active imagination inclined toward outlines rather than vivid sensory details—and a strong tendency toward humorous self-analysis (Wells 1934; Delbanco 1990; MacKenzie and MacKenzie 1973). Taken together, this suggests a rather uncommon personality profile that goes some way toward explaining the paradoxical impression he has left on scholars: imaginative without being neurotic, social without being agreeable, assertive and ambitious without being narcissistic, open to experience but with an extreme need for order and goal-orientation. This was the temperament that made Wells act as a scientist or journalist among literary people and as an imaginative writer among those who invoked science against him, shamelessly and humorously shifting the goal posts.

Wells's personality profile can be illuminated by a comparison with the neurotic, aesthetic, narcissistically inclined Jack London. The two authors' life stories are remarkably similar in external particulars. They were raised in the lower middle class, were influenced by Darwinism during their intellectually formative years, and rose to literary prominence via magazine publications (Wells 1934; Haley 2010). Both were eclectic socialists accused of intellectual and ideological inconsistency. Both also became hugely successful and influential authors of fantastical adventure stories. But Wells finally rejected the extreme individualism of Spencer and Nietzsche that seduced London even as he attempted to satirize it. London produced Ernest Everhard, and Wells produced semi-satirical socialist "heroes" who bungled their way through versions of his own life (1899; 1978 [1900]). Orwell accused Wells of being "too sane" and of having "lost all power of action" in the same essay that he accused London of having "a fascist streak" (1941). London was consumed by aesthetic ambition, and wrote an admiring letter to Joseph Conrad. Wells wrote satirical novels against aesthetic impressionism and argued about it with both Conrad and James (1915; 2008 [1932]; 1934, 410–24;

Delbanco 1990). Wells planned thousands of years into the future and turned short-term happiness into the great antagonist of humanity, and London declared that he would “rather be ashes than dust” (Shepard 1956, viii). Ultimately, London watched the stars and saw conflict at the core of the universe—a cosmos of clashing systems and screaming planets. Wells watched the same stars and saw stable laws communicating “infinite peace and protection” (1978 [1896], 156).

This chapter outlines Wells's worldview, tracing its roots in the psychological challenges of Darwinism and its transmutations through his personality and life experiences. I discuss Wells's position as a founder of modern science fiction, and the particular narrative techniques of his scientific romances. Finally, I analyze two of his most famous science fiction novels: *The Time Machine* (1895) and *The Island of Doctor Moreau* (1896). These two novels have remained in the public consciousness and have together had at least ten cinematic adaptations between 1913 and 2002. Yet, they represent two very personal modes in which Wells dealt with naturalistic horror. They are nightmares of his evolutionary mythology.

6.1 The Little Animals and the Mind of the Race

Wells's worldview was founded on evolutionary biology. He called his year in T. H. Huxley's biology class “beyond all question, the most educational year of my life” (1934, 161). His first published book was the slim *Textbook of Biology* (1893), and his last—appearing in the year of his death—was a reflection on human extinction written “under the urgency of a scientific training”: *Mind at the End of Its Tether* (1946, 1). In this final proclamation, largely a concession of “stoical cynicism” in the face of uncertain futures, Wells retains his foundation: “One thing is certain. Not one fact has ever emerged, in a stupendous accumulation of facts, to throw a shadow of doubt upon what is still called the ‘Theory’ of organic evolution” (1946, 29, 30). Yet, the *Mind* whose end he predicts is a quasi-spiritual human hive mind. It echoes “the Mind of the Race,” his most grandiose description of human culture (1915, 168, 169).

For Wells, biology shaped social theory, and social theory shaped aesthetics. But nature and culture were split all the way down. His social theory was premised on his idea that culture is an "artificial process," integral to human evolution but opposed to "natural man" (1975 [1896]; 1975 [1897]. He believed that culture could fundamentally refashion human psychology. It could be "moral padding" that "fit the round Paleolithic man in the square hole of the civilized state" (1975 [1897], 221). Because of this potential for cultural reprogramming, the World State should be pursued without regard for the natural limitations of humanity: "moral theory should be deduced from an ideal social state" (1975 [1897], 227). Culture was itself shaped by anyone who contributed to public life: scientists, philosophers, priests, politicians, and artists (1975 [1897], 218). From this perspective, the role of the responsible novelist is clear. Wells set out to imagine the ideal social state and to educate his readers toward it. But in order to do so, he suppressed his intuitive understanding that culture must inspire as well as restrain emotion, and he forgot that "natural man" has prosocial dispositions.

Wells's division between nature and culture is a variant of a common form of moral and social dualism. His early theoretical essays describe "natural man" as egocentric and violent, and "artificial man" or "social man" as the cultural restraint on those impulses (1975 [1896]; 1975 [1897]). He retains this basic division in one of his last books, where he describes humans as "a primarily solitary and ego-centered species" that was "forced" to cooperate by the invention of language during the Paleolithic (1995 [1942], 59). These passages suggest an image of human nature as a vicious beast restrained by an external force. The image would not have been out of place among Robert Louis Stevenson's Christian reflections. It also echoes Wells's enduring fascination with Plato (1934, 394). Along with Huxley, Wells continued the tradition of pitching egocentric emotion against rational restraint. Indeed, Wells claims that the beast would return to its ways if cultural restraints were removed: human mental advances of the last 50,000 years would be swiftly "reversed" (1995 [1942], 60). In this bare-bones account, he had simply provided a Darwinian guise for the ancient idea of nature as a threat to transcendental reason.

Though Wells stuck stubbornly to his dualistic theory of human evolution, his idea of nature was more complex than his pronouncements on "natural man." His study of biology gave him a broad sense of the ingenuities of life, a scalpel's touch to and a microscope's view of the natural world, and a range of particular impressions of non-human animals. That perspective induced enough humorous respect that, as an autograph in his

Textbook of Biology, he sketched himself being dissected by a giant rabbit (Collinge 2017, 49; Library of Illinois). Like Huxley and Jack London, he locally acknowledged the survival value of cooperation, even for species without culture (1975 [1892]; 1975 [1895]a).

Despite Wells's suspicions against natural man, *Anticipations* contains an interpretation of nature that matches Ernst Haeckel's equation of the ecosystem to God. Wells's version, unlike Haeckel's, is optimistic—a jaunty form of Might is Right, culminating in human exceptionalism:

> The insoluble problems of pain and death, gaunt, incomprehensible facts as they were, fall into place in the gigantic order that evolution unfolds. All things are integral in the mighty scheme, the slain builds up the slayer, the wolf grooms the horse into swiftness, and the tiger calls for wisdom and courage out of man. All things are integral, but it has been left for men to be consciously integral, to take, at last, a share in the process, to have wills that have caught a harmony with the universal will, as sand grains flash into splendour under the blaze of the sun. (1902, 292)

These claims seem contrary to the idea of a vicious natural man who needs restraining. Suddenly, nature is a divine order and human minds are a conscious expression of its divinity. In the same passage of *Anticipations*—concerned with the future "religion" of the World State—Wells declares that if the universe appears "non-ethical by our present standards," we must "reconsider these standards and reconstruct our ethics" (1902, 286). In other words, ethics should be shaped by nature. This is a direct contradiction of his idea that moral theory should derive from "an ideal social state" (1975 [1897], 227). Nevertheless, even in this Panglossian account of nature, the final emphasis remains on culture. It is by *understanding evolution* that humans have become conscious. It is through their scientific consciousness that humans harmonize with nature's "will" and become more than Little Animals. Wells's "flash of splendour" from a divine sun reflects the light of a culture that Huxley imagined emerging from "the infinite source of truth" (Huxley 2009 [1863], 112).

Wells's theory of the world made the harsher side of nature tolerable for him. Sometimes, he imagined human reason as the culmination of nature. Sometimes, he imagined the two forces locked in a struggle. In one scenario, he was part of a righteous fate; in the other, he was fighting a righteous war. His quasi-religious vision of nature conflated scientific comprehensibility with divine intention. Either the universe is "one and systematic, *and held together by some omnipresent quality*" or it is simply "a casual aggregation, an incoherent accumulation with no unity whatsoever outside the unity of the personality regarding it" (1902, 281, my italics). As he put it: "to believe the former is, to any one not too anxious to quibble, to believe in God" (1902, 281). This way, the cosmos appeared to grant a simple choice: muddle or divinity, unconsciousness or science, little animals or the Mind of the Race. And once the choice was made, there was peace in the righteous struggle.

As a young man, Wells achieved a unification of the universe that resembles Martin Eden's discovery of science: "a realization that the synthesis of the sciences composed a vital interpretation of the world" (1934, 160). But unlike Martin (and London), Wells did not descend into suicidal depression as a result. He was spared some cosmic suffering because of his equable personality—he was not prone to recurrent episodes of depression. But his idea of the World State also articulated and supported his temperamental equanimity. He could believe that he saw the world whole because he had split it so studiously, and because he adhered to his theory so stubbornly. In the latter phase of his life, "one wakeful night," he saw his utopian vision dispassionately: "a helpful dream that kept me going along some of the more exacting stretches of the course, a useful but essentially an impossible dream" (1934, 8, 9). His autobiography begins with this late-night scribble, but frames it as a momentary weakness, and reasserts the World State as the purpose of his life: "the crystal I have brought out of solution" (1934, 12).

Wells's confidence in the World State made him stand out among his contemporaries. His teacher Huxley imagined culture as a fragile and ephemeral triumph. Darwin had only a flickering notion that social instincts might expand across the globe. Edward Bulwer-Lytton created evolved utopian monsters. Jack London turned away from biology to protect his socialist dream. Wells took the same Darwinian ideas and shaped them into a modern utopia for which he preached—albeit with varying levels of optimism—for half a century. Unlike Stevenson, who saw a divine struggle between Good and Evil in the human soul, Wells believed that the struggle could be resolved by human hands. He saw the World State as a possibility until he despaired of it in his final book (Earle 1950; Wells 1915). Often, as in *Anticipations*, he portrayed the World State as inevitable. But his attitude toward it, even during his final moment of disillusionment, consistently had the character of religious faith.

Ten years before his death, Wells claimed that he had never wholly left his "sturdy atheism," but two years later, he called himself an "essentially religious person" (Wells, 1934, 578; 2008, 222). He went through an overtly religious period earlier in his life (1917; 1919; 1934, 577). When he denounces this religious period in his autobiography, he also approvingly cites a dialog from his novel *The Undying Fire* (1919) in which the narrator asks God "Am I right to fight?" and is answered that Mind transcends matter: "So long as your courage endures you will conquer ... By courage it is that the stars continue in their courses, day by day. It is the courage of life alone that keeps sky and earth apart" (1934, 577). This passage, written as a culmination of Wells's relationship with God, resonates with other impressions of cosmic order and intention throughout his career. Sometimes his narrators find peace in the mere existence of cosmic laws, which Wells merged with cosmic intention. The Time Traveller from his first novel momentarily finds "all the gravities of terrestrial life" dwarfed by the slow movement of the constellations (2005 [1895], 45). Prendick, the unlucky visitor to *The Island of Doctor Moreau,* clings to stargazing after his return to civilization, finding solace for "whatever is more than animal in us" in "the eternal laws of matter" (1978 [1896], 156). At other times, the cosmic impressions are directly linked to the Mind of the Race. In the apocalyptic short story "The Star" (1897), a scientist faces the earth-threatening comet with a courageous appeal to rational consciousness:

> He looked at it as one might look into the eyes of a brave enemy. "You may kill me," he said after a silence. "But I can hold you—and all the universe for that matter—in the grip of this little brain. I would not change. Even now." (1983 [1897], 226)

This declaration may as well have been directed at the cosmos itself, or at nature. For the leader of the World State in *Things to Come*, toward the end of Wells's career, the stars become the ultimate counter to the idea that humans are "little animals" (1936). But two decades earlier, Wells had satirized his alter ego Boon for describing the literature and the Mind of the Race with "swinging reference to the stars" (1915, 175).

The Mind of the Race—Wells's idea of culture—was a construct of faith, riddled with logical inconsistencies and personal ambivalence. On the one hand, Wellsian *culture* centered on emotional restraint, science, and self-transcending rationality. He described his persona as "a preoccupied mind devoted to an exalted and spacious task and seeking a maximum of detachment from the cares of this world and from baser needs and urgencies" (1934, 9). Orwell could plausibly claim that "History as he sees it is a series of victories won by the scientific man over the romantic man" (1941). On the other hand, Wells's personal and literary expression was fuelled by imaginative culture. His emotional experiences with art were not just the distractions of natural man: they helped him endure harsh conditions and imagine new futures—for himself, and for humanity generally. As a literary writer and amateur cartoonist, he lived by evoking and directing emotion. He locally admitted that "social man" was shaped by those who inspired as well as restrained emotion: "eccentric and innovating people, playwrights, novelists, preachers, poets" (1975 [1897], 218). His final suppression of this truth, and his persistent dichotomizing of emotion and moral education, produced a deep ambivalence in his attitude to literature.

Imaginative culture was important to Wells throughout his life. He read fiction from the age of seven and began drawing at eleven or twelve, "spontaneously and expressively" (1934, 15, 53). As a young man, on his way to drop out of the Normal School of Science, he found direction by inquiring into what "such exalted names as Goethe and Carlyle, Shelley and Tennyson, Shakespeare, Dryden, Milton, Pope—or again Buddha, Mahomet and Confucius—had had to say about the world" (1934, 194). In his second marriage, he formed the habit of "making endless doggerel" in the morning and abandoning work in the evening to draw "silly little sketches about this or that incident which became at last a sort of burlesque diary of our lives" (1934, 365). His scientific romances were influenced by Jules Verne, and he leaned heavily on Hawthorne for his first draft of *The Time Machine* (1934, 253, 254). He absorbed Plato and William Morris, fashioning *The Time Machine* and *A Modern Utopia* partly in response to their fictional societies (1934, 394). His imaginative debt to satirists like Voltaire and Swift is apparent both in his literary tone and in the style he adopts in letters. At one point, he openly regretted having addressed a friend in the manner of a "Cockney Voltaire" (1934, 305). These aspects of Wells form a persistent undercurrent to his emotionally detached "persona." While analyzing his self-image, he acknowledged this undercurrent, but displayed it in a negative light: as "distractions, attacks and frustrations," as "antagonisms and temptations" (1934, 10). Ultimately, here as elsewhere, he gave precedence to self-transcending rationality: "My *persona* may be an exaggeration of one aspect of my being, but I believe it is a ruling aspect" (1934, 11).

Wells's tendency toward imaginative play gave him strong intuitions about the power of imaginative culture. He suggests that his reading of novels as a child shaped the course of his life—even, he speculates, made the difference between being "alive to-day and writing this autobiography" and "being a worn-out, dismissed and already dead shop assistant" (1934, 53). When he reflects on his narrative poems and cartoons, he calls them "silly little sketches" and "apparently very trivial things"—but he also observes that they were "really very important" for his marriage (1934, 365):

> Into this parallel world of burlesque and fancy, we transferred a very considerable amount of our every-day life, and there it lost its weight and irksomeness. We transferred our own selves there also. Miss Bits became an active practical imperious little being and Mr. Bins a rather bad, evasive character who went in great awe of her. He was frequently chastised with an umbrella or "Umbler pop." . . . These may seem at the first glance to be the most idle of scribblings but in fact they are acute statements in personal interpretation. (1934, 366)

The parallel world of fancy was a way for Wells and his wife to work out their relationship to each other, to their own identities, and to their rapidly changing social circumstances as Wells rose to fame. Their various life events—publications, house fires, life-threatening illnesses, and trips abroad—were transposed into a weave of fiction that made life more manageable by organizing it into a system of evocative symbols. The parallel world allowed them, as Wells put it, to find their "*modus vivendi*" (1934, 365). Had he translated those astute observations into the language of his theoretical worldview, he might have acknowledged that similar "apparently very trivial things" affect the modus vivendi of all humans. The director of the World State was clearly not on speaking terms with Miss Bits and Mr. Bins.

Wells's stubborn dichotomy between natural and artificial man also split literature in half. His conflicted reflections on his own profession contain two main interpretations: literature as trivial distraction, and literature as moral education toward the World State (1915, 133–77; 1934, 410–24). Seen purely as art, a novel is like "a market place or boulevard": a more or less attractive background that you pass by "on your various occasions" (1934, 411). Seen as moral education, literature is part of the Mind of the Race: the "great vague body of man's super-personal mental life, his unselfish life, his growing life, as a premeditating, self-conscious race and destiny" (1915, 168, 169). Wells was most preoccupied with these two versions of literature when he discussed Henry James and Joseph Conrad—two authors of distressingly greater artistic talent, with whom he had fierce debates, and whose novels did not primarily aim to be moral education (1934, 410–19). But Wells also distinguished between his own trivial novels and his serious ones.

As Wells saw it, his trivial literary works were his early science fiction novels: "playful parables" that aimed primarily to "intensify our natural reactions of wonder, fear or perplexity"; his serious works were his later social realist novels, which aimed to be "ethical inquir[ies]" about "problems of adjustment" (1978 [1934], iii, vi; 1934, 410, 411). These comments illustrate both Wells's attitude to his own fiction and the particular distortions inherent in his worldview. He saw human development clearly enough to predict tank warfare and Wikipedia, suburban spread, and the atom bomb ("The Land Ironclads" 1903; *World Brain* 1938; *Anticipations* 1902; *The World Set Free* 1914). Yet, he consistently underestimated the power of symbols and emotions to sway people, transport them, and inspire actions that went beyond "purely personal particulars" (1915, 171). Because of this failure, his evaluative judgment about his own fiction is diametrically opposed to the judgment of history.

6.2 The Shakespeare of Science Fiction

Though science has entered into fiction as long as it has entered into life, Wells can be safely described as one of the founders of modern science fiction (Aldiss 1986; Alkon 2016; Rabkin 1983). He brought biology—and specifically Darwinian evolution—definitively onto the literary stage. He introduced an element of bold speculation, even semi-magical extrapolation, that caused conflict with the cautiously steam-powered Jules Verne (Asimov 1968). And with his "genius for plot," in Harold Bloom's phrase, he gripped and held a wide audience that remembered his stories (2005, xii). Darko Suvin's influential description of *science fiction* as "cognitive estrangement"—a plausible novelty introduced into a realistic framework—might be mistaken for a summary of *The Time Machine* (1977). Wells's scientific romances spoke to his contemporaries "in a series of dreamlike images, which distil the paradoxical hopes and fears of his age" (Aldiss 1986, 145). But they speak far beyond his age.

Many people have speculated about the narrative formula behind Wells's success. Henry James marveled at the "spell" of Wellsian storytelling, which for him was an experience of Wells's own character, overshadowing objections to the craftsmanship (1934, 413). Isaac Asimov commended Wells for "disposing of his gadgetry quickly, with a slash of fantasy reduced to plausibility" to "deal with *human beings*" (1968, 16). Jules Verne complained that *The First Men in the Moon* had been successful despite the feigned scientific legitimacy of its gravity-insulated space craft (quoted in Asimov 1968, 15, 16). Recent critics have accused Wells of using "misleading layers of purported accuracy and control, grounded in public knowledge... to disarm the incredulity of readers", or they have simply explained his successful storytelling as an ability to "formulate 'some of his serious speculations' in a digestible form" (Glendening 2002, 571; McLean 2009, 1).

Though interpretations and evaluations differ among critics, one basic impression seems to be that Wells was a particularly adept illusionist, directing the eyes of his audience away from pedals, strings, and hidden cards. That impression accords with Wells's own characterization of his early fantastical stories as "good gripping dreams," sweeping readers along despite their better judgment (1978 [1934], iii). The illusions evidently worked even when the audience was peripherally aware of the strings. Something in Wells's storytelling made people like Henry James sit through the show even as they mouthed complaints about its quality:

> I always read you … as I read no one else, with a complete abdication of all those "principles of criticism," … with which I roam, with which I totter, through the pages of others … For under your spell I do advance—save when I pull myself up stock still in order not to break it even with so much as the breath of appreciation … (quoted in Wells 1934, 413)

The narrative structures of Wells's early fiction were simple but effective. They built on literary traditions of satire, utopian vision, adventure, war-stories, and—in accordance with Wells's habit of writing short impressionistic essays for publication—anecdote. Wells used a plethora of basic psychological tricks to stimulate his readers and align their perspectives with his own. He wrote in a conversational voice approximating letters to his friends (1934, 253). He frequently took readers into his confidence through direct address, inviting them to share in a joke or a speculation about the credibility of the story he was telling (see the narrator of *The Time Machine*, the introduction to *Doctor Moreau*, "The Plattner Story," "The Purple Pilcus"). He set up simple twists and contrasts: A war correspondent's confident expectation of the battle outcome increased the surprise of his "Land Ironclads," and dismissal of perfect flying machines as the "easy dreaming of a literary man" gave credibility to the gritty struggle of his "Argonauts of the Air" (Wells 1975, 86, 218). He offered narrative perspectives that did not insult the reader's intelligence, did not appear to think too much of themselves, and did not demand very much of the reader. His participant narrators tended to be sensible everymen—often intelligent, but unable to fully describe the fantastic events that took place, and confused by "what [the scientist] was pleased to call elementary mathematics" (Wells 1978, 396). When he used an omniscient narrative voice, he offset the godlike perspective with frequent free-indirect-discourse forays into his characters' limited perspectives (e.g. *The Invisible Man*, "The Star"). In the world of simulated social interaction that is literature, everything Wells did in his early fiction was geared toward making the reader his friend.

The stories' inviting narration was enhanced by the allure of the fantastic. Wells's early fiction was often built on speculation about scientific concepts or extraordinary discoveries. "Davidson's Eyes" and "The Plattner Story" spring from the idea of time as a fourth dimension; *The Time Machine* adds geological and evolutionary speculations to that notion; and "The Grisly Folk" was conjured from Neanderthal bones. McLean admirably traces how Wells's fiction entered into contemporary scientific debates (2009). But although Wells's scientific literacy is important, it was just one source of the fantastic element that made his early fiction successful. His stories extrapolated from supernatural concepts as well as scientific ones—as in "The Man Who Could Work Miracles," "The Lord of the Dynamos" and "Jimmy Goggles the God"—or used a thin varnish of science to make fantasy look plausible, as in *The Wonderful Visit* or *In the Days of the Comet*. The early fiction often outlives the scientific concepts from which it was derived, like the vivisection in *Doctor Moreau*. Its enduring appeal, or its main appeal for Wells's contemporary readers, went deeper than scientific discourse.

Wells understood that he needed to violate the perceptual biases of his readers to just the right degree. As he put it, his fantastic notions were made interesting by "their translation into commonplace terms and a rigid exclusion of other marvels from the story"—because "nothing remains interesting where anything may happen" (1978 [1934], iv). Twentieth-century science fiction theory dubbed this idea *novum*: a plausible novelty introduced to a realistic setting (Suvin 1977). More recently, reader response experiments have shown that stories are more memorable if they incorporate a few counterintuitive elements into a plausible, meaningful whole (Norenzayan et al. 2006; Upal 2011). Wells launched his early success using this principle, building on old fantastical genres like fairy-tale and religious allegory. In *The Invisible Man*, the titular character has to deal with a realistic world of hunger, cold, curious or suspicious rural people, and inconveniently visible footsteps. The reader is carried along under the guise of plausibility to imagine just what it would be like to be invisible, which advantages and disadvantages there might be, and how one might deal with an invisible foe. On a more general level, the novel deals with the integration of new technological advantages into society, and the social responsibilities of a genius, or the effects of research obsessions carried out in isolation. Many of Wells's counterintuitive concepts are set in similarly realistic settings as integral parts of a storyline—such as the anti-gravity alloy of *First Men in the Moon*, the beast people of *The Island of Doctor Moreau*, the humanoid fish of "In the Abyss," and the deified dynamo of "The Lord of the Dynamos."

The focus on fantastical events seldom left room for in-depth exploration of inner lives. To Harold Bloom, Wells was "the least Shakespearean of all English writers"—a "tale-spinner" more concerned with "plot" than with the human condition (2005, xii). But intuitions about humanity can take different forms. Though Wells did not tend to fashion his characters as fully-rounded persons, he fashioned much of his early fiction with an acute understanding of the persons on whom his fortunes depended. He flattered his readers, talked straight to them, teased them, made them laugh, and gave them stimulating puzzles in familiar frames. The strings and pedals of his illusions do show at times—in the seeming afterthought of including the protagonist's wife in *The War of the Worlds*, for example, or in the tiresome parade of insignificant colorful characters in the beginning of *The Invisible Man*. But innumerable readers have been swept along under the spell. When Wells was at his best in the early fiction, he combined his ability to effectively manipulate emotion with his imaginative impression of the human condition—a wondering mind trapped in animal meat, molded through deep history by unforgiving laws, carrying a fire that could either light the way to the stars or turn the moment into ashes.

Because of Wells's theoretical and psychological ambivalence, his fantastical stories portray culture in contradictory ways. In *The First Men in the Moon* (1901), it takes the shape of the Grand Lunar—a brain surrounded by living encyclopedias that controls the perfectly organized, rational society on the moon. In "A Story of the Stone Age" (1897), it appears as emotional images: dreams or half-perceived suggestions, anthropomorphisms of animals and the dead, guiding the primitive people to new ideas and new interpretations of the world. In *The Wonderful Visit* (1895), it is a parallel dimension full of beauty and mythical creatures—a realm without pain or passion, which sometimes sends impressions through the dimensional veil to those who are asleep, but cannot be appreciated by the man of science. To the Time Traveller lamenting its loss in the distant future, it is *The Philosophical Transactions of the Royal Society* and his own "seventeen papers upon physical optics" (2005 [1895], 49). One aspect of culture is a rational pursuit of knowledge, the other a mysterious realm of escapism and individual wish fulfillment. Wells rarely connected these two versions of culture even in his fiction. "A Story of the Stone Age," which portrays a culture that both strives forward intellectually and appeals to emotions, is a rarity. In line with his differentiation in literature between ethical inquiries and playful parables, he tended to portray the rational progress and the imaginative escapism as distinct absolutes.

Wells's grand symbolic visions of humanity tended to depict culture in its rationally progressive guise. The stories that center on culture as an indulgence were often themselves indulgences. *The Sea Lady* (1902) summons a mermaid to sate Wells's "craving for some lovelier experience than life" and to tickle his occasional longing for "impossible magic islands" (1934, 393). "The Door in the Wall" (1906) tells of a man who finds the gateway to a magical garden as a boy and pines for it throughout his life, though he always chooses to forsake it for worldly matters when it appears. There is some symbolic power in both these tales, but it might not be unjust to call them playful parables. Their main theme is personal escapism. In contrast, there is the coldly reasoning superiority of the Martians from *The War of the Worlds*, the logical expediency of the

Selenites from *The First Men in the Moon*, and the scientific perspective of *Time Machine* and *Moreau*. These novels deal with the fragile nature of human uniqueness, the precarious psychological and material conditions of humanity. Though they were themselves instances of imaginative culture, offering relief and direction to their author and readers, they did not include imaginative culture as a factor in their fictional universes. Nonetheless, the best of them register a friction between their image of human evolution and the reality of human life. *The Time Machine* (1895) and *The Island of Dr Moreau* (1896) are prominent examples.

6.3 *The Time Machine* (1895)

The Time Machine is an extrapolation from the class system of Wells's contemporary society. It depicts a world eight-hundred-thousand years into the future, where "Capitalists" and "Labourers" have become two distinct humanoid species: the surface-dwelling Eloi who live in ease, and the subterranean Morlocks who keep the Eloi as cattle, tending to them and feeding on them (2005 [1895], 36). The story of this future is told to the novel's narrator by his friend, who is only identified as "the Time Traveller." The novel starts with the Time Traveller's revelation to a circle of friends that he has discovered time travel. During a dinner a week later, he returns from a journey into the future, and relates his experiences with the Eloi and Morlocks. In his story, he arrives and meets the Eloi, loses his time machine, rescues a drowning Eloi girl named Weena, explores the surface society, discovers the Morlocks' underground society and realizes that they are feeding on the Eloi, escapes a wildfire, loses Weena, is attacked by the Morlocks, regains his time machine, and travels forward in time to escape. Before returning to his own time, he witnesses the final gasps of life on earth under the dying sun. Throughout the story, the narrator acts as a sensible everyman, doubting the veracity of the Time Traveller's fantastical claims, but respecting his scientific acumen and listening with an open mind. The novel ends with the narrator visiting the Time Traveller's home again, and catching a glimpse of him departing on another journey through time—from which, we are told, he does not return.

The Eloi and Morlocks together represent the nightmarish consequences of humanity's failure to pursue Wells's global scheme. Their ancestors were not guided by the long-term perspective of the World State. Instead of planning for the natural and artificial evolution of humanity, they set themselves "steadfastly toward comfort and ease, a balanced society with security and permanency as its watchword" (2005 [5], 57). But the novel is not didactic in any simple sense. The Time Traveller discovers the nightmare gradually and interprets it provisionally. At first, he sees only the frail, angel-featured Eloi, and supposes that their mental feebleness is a result of utopian comforts. When he discovers the pale, ape-like Morlocks, he realizes that the Eloi are their cattle. Though the Time Traveller's final interpretation is that the human intellect has "committed suicide" by aiming for short-sighted comforts, he emphasizes that it is only a hypothesis: "It is how the thing shaped itself to me, and as that I give it to you" (57). He never articulates Wells's antidote to the catastrophe. When he escapes from the Morlocks into the further future, his view of the dying sun over rudimentary organisms suggests that all social antidotes are moot. When he disappears on his final journey, he leaves the last word to the narrator, who expresses a weak optimism and declares that the Time Traveller is a pessimist (66). Readers are left to contemplate the horrific fate of humanity, the ultimate end of life on earth, and the possible attitudes that might make such contemplation tolerable.

The dreamlike depiction of geological time is one of *The Time Machine*'s enduring achievements. With the simple conceit of describing time as a "fourth dimension," Wells makes his readers consider it anew, as if they were two-dimensional creatures collating scattered impressions into the concept of a third dimension (3, 4). The Time Traveller struggles to convey to his dinner guests the sensation of being propelled through time, a nauseating feeling like a "helpless headlong motion" (14). His description of the journey proceeds in lyrical particulars from days to seasons to the passing of human civilizations, until the earth itself flows like a river under his machine:

> As I put on pace, night followed day like the flapping of a black wing. The dim suggestion of the laboratory seemed presently to fall away from me … the palpitation of night and day merged into one continuous greyness; the sky took on a wonderful deepness of blue, a splendid luminous colour like that of early twilight; the jerking sun became a streak of fire, a brilliant arch, in space … I saw trees growing and changing like puffs of vapour, now brown, now green; they grew, spread, shivered, and passed away. I saw huge buildings rise up faint and fair, and pass like dreams. The whole

> surface of the earth seemed changed—melting and flowing under my eyes. (14, 15)

These passages bring readers gradually from the safety of a Victorian "smoking-room"—with its comfortable chairs, cigars, and "bright circle of the little lamp"—to a cosmic distance that transforms the shape of the sun and earth (12, 13). Wells's spell is deftly cast, grounding the inhuman scope in details that allow a human mind to visualize and extrapolate.

Once the Time Traveller arrives in the future, the focus shifts from cosmic time to Wells's core concern: the fate of humanity. The traveller's first sight of the year 802,701 is the face of a decaying marble sphinx. That sphinx is a classical allusion worthy of Wells's first draft of the story: the *Chronic Argonauts* (though he later lamented "the ineptitude of this rococo title for a hard mathematical invention" (1934, 254). Like its mythological counterpart, the Time Traveller's sphinx poses a riddle about man. Its marble shape is tied to both species of human descendants. It acts as a hiding place for the Morlocks (who stir inside of it with "a chuckle"), a place where they store the stolen time machine, a trap they set for the Time Traveller, and a terrifying symbol of them to the Eloi (28, 57, 58). It frames the Time Traveller's first sight of the Eloi: "Coming through the bushes by the White Sphinx were the heads and shoulders of men running … perhaps four feet high" (17). It forms the background to his bond with the Eloi girl Weena: "the little doll of a creature presently gave my return to the neighborhood of the White Sphinx almost the feeling of coming home" (32). The sphinx's human-like face, at first hazy through a thunderstorm, haunts the Time Traveller throughout his visit: "the sightless eyes seemed to watch me; there was the faint shadow of a smile on the lips"; "the sphinx of white marble, which had seemed to watch me all the while with a smile at my astonishment"; "white, shining, leprous... it seemed to smile in mockery of my dismay"; (16, 19, 26). Appropriately, this sphinx also adorned the cover of the first edition of *The Time Machine*. Though Wells named his story after the machine that rushes past civilizations and species, the novel teases endlessly at the riddle of humanity.

Wells's split view of human evolution comes through in his depiction of the Eloi and Morlocks. Both species have lost the capacity for cumulative culture, and therefore the qualities Wells associates with "artificial man": community, self-transcendence, fellow-feeling, and rational consciousness. The Eloi are frail dolls, beautiful and hapless. They have a rudimentary language, display no signs of joint effort, engage in no educational enterprises, and seem to have no shared culture beyond the language: "They spent all their time in playing gently, in bathing in the river, in making love in a half-playful fashion, in eating fruit and sleeping" (31). They have become a slothful extreme of human decadence, taking "to art and to eroticism"—but even their limited art is fading:

> To adorn themselves with flowers, to dance, to sing in the sunlight: so much was left of the artistic spirit, and no more. Even that would fade in the end into a contented inactivity. (25)

The Eloi's similar dress and near sexless bodies initially give the Time Traveller the impression that they have achieved "Communism" (21). But he soon finds out they have no sense of community. Because they lack the ability to maintain "interest" in any particular event, they lack conscious self-interest, but they also lack any sustained interest in each other (21). The Eloi appear to be socially and sexually equal, but their species is effectively dissolved into isolated individuals. We get a full demonstration of their inhumanity when a group stands by complacently to watch Weena drown (31).

The Morlocks are given less descriptive room than the Eloi, but when they appear, they are monstrous. Their physical appearance puts them on a level with the Missin' Links of *The Lost World:* "bleached, obscene, nocturnal" things, "human rats" with "pale, chinless faces," "great, lidless, pinkish-grey eyes," "flaxen hair" down their backs and "a peculiar unpleasant odour" (34, 40, 41, 54). They hunch like apes, and the first that appears clambers into the underground "like a human spider!" (34). These features don't just make them beastly but "nauseatingly inhuman"—the more disturbing because they are "inhuman sons of men" (41, 45). The Time Traveller occasionally pities them, as when he watches them dash blindly into the forest fire he has started (55). They appear to have a language, and they show signs of coordinated intelligence when they set a trap with the time machine in the sphinx (39, 40, 58). But they never emerge as rational beings. Even their conquest of the Eloi has the character of animal unconsciousness:

> The Morlocks made [the Eloi's] garments, I inferred, and maintained them in their habitual needs, perhaps through the survival of an old habit of service. They did it as a standing horse paws with his foot, or as a man

> enjoys killing animals in sport: because ancient and departed necessities had impressed it on the organism. (42)

To the Time Traveller, the Morlocks are primarily a predatory force. They fill the darkness, pawing at him, hiding his machine, and dragging the Eloi underground for meat. His only visit to their society reveals nightmare corridors filled with thumping machines, "very stuffy and oppressive," with "the faint halitus of freshly-shed blood" (40). Before his weeklong visit is over, he has made up his mind about these descendants: "I longed very much to kill a Morlock or so … it was impossible, somehow, to feel any humanity in the things" (49).

Both Eloi and Morlocks enhance Wells's horror vision. Though the Time Traveller tries to maintain "a scientific spirit," he oscillates between pity and disgust for his descendants (46). His antipathy for the Morlocks is evoked partly by sympathetic identification with the Eloi (45, 46). But the Eloi have, in a certain light, earned their Morlockian fate: "thousands of generations ago, man had thrust his brother man out of the ease and the sunshine … now that brother was coming back—changed!" (42). The Time Traveller speaks of this as the Eloi's "Nemesis"—a goddess of vengeance to complement the riddling sphinx (42). The Eloi's aristocrat ancestors are implicitly to blame for the degeneration of humanity as well as for its divergence into two species: they were the ones who killed the human intellect by aiming only for "a balanced society with security and permanency as its watchword" (57). The Morlocks horrify the Time Traveller with their brute predation, but the Eloi infuriate him with their indolence and incompetence (27, 28, 31). The Eloi appeal more strongly to his sympathies mainly because they have "kept too much of the human form" (46). His ambivalent pity for both species elevates the novel above the adventure frolics of Arthur Conan Doyle. It strengthens the emotional link between the nightmare future and humanity's present.

The Time Traveller's experience is an indictment of late-Victorian political thinkers across the spectrum: conservatives who sought to stabilize the class system and socialists who sought to create secure, permanent utopias. But there is also a hint of Wells's concern for the consequences of his own planned society. He wanted peace and efficiency but was preoccupied with the enfeebling effects of domestication. The Time Traveller lingers at some length on this apparent paradox:

> The ruddy sunset set me thinking of the sunset of mankind. For the first time I began to realise an odd consequence of the social effort in which we are at present engaged. And yet, come to think, it is a logical consequence enough. Strength is the outcome of need; security sets a premium on feebleness. The work of ameliorating the conditions of life—the true civilising process that makes life more and more secure—had gone steadily on to a climax. One triumph of a united humanity over Nature had followed another. Things that are now mere dreams had become projects deliberately put in hand and carried forward. And the harvest was what I saw! (23, 24)

Though this is an early hypothesis, the Time Traveller's discovery of the Morlocks only changes his idea that humanity was "united." He ultimately believes that the two classes were secure—and hence enfeebled—in their separate spheres: "The rich had been assured of his wealth and comfort, the toiler assured of his life and work … And a great quiet had followed" (57). Civilization itself enfeebles. In this view, the degeneration of humanity appears more like fate than suicide.

The temperamentally optimistic Wells did not bow to fate. To him, the paradox of civilization was not a tragedy, but a problem to solve. He faced that problem head on during the 1890s. The same year as he published *The Time Machine*, he spoke with the Time Traveller's voice in his critical review "Bio-Optimism":

> a non-combatant species, if such a thing as a non-combatant species ever exist, a species, that is to say, perfectly adapted to static conditions, is, by virtue of its perfect reactions, a mechanism ... In brief, a static species is mechanical, an evolving species suffering—no line of escape from that *impasse* has as yet presented itself. (208–9)

Stability and security seem directly opposed to the Mind of the Race. The Eloi and Morlocks have become inhuman—feeble mechanisms actuated by habit—because they have been perfectly adapted to unchanging peace. The "impasse" behind this vision is simplistic. Wells implies a number of erroneous ideas: that adaptation to static conditions eliminates the struggle for survival, that rigidly instinct-driven species cannot evolve rapidly, and that the members of a rapidly evolving species suffer more than those of a relatively static species. Nevertheless, he had a keen sense of the basic logic of evolution. His "impasse" previsions current hypotheses that the human niche—with its behavioral flexibility and cumulative culture—arose as a result of rapidly changing environments (Hill et al. 2009; Richerson et al. 2008). In the childlike Eloi, he captures the *neoteny* effect now known to be a feature of domestication: the tendency to retain juvenile characteristics into adulthood (Bjorklund 1997). And his ideas about the effects of security align with contemporary studies of human self-domestication (Henrich 2015; Wrangham 2019).

Wells's solution to his evolutionary impasse was science. *Anticipations* paint a planned society that constantly challenges its inhabitants—toward greater organization, concerted exploration, and conscious understanding of the natural world. In his vision of the World State some thirty years later, Wells explicitly advocates a society that does "not abolish death or danger" but instead makes "death and danger worthwhile" through scientific exploration (1936). As he put it, the world "will never be safe for men, and there is no happiness in safety" (1936). *The Time Machine* only contains this solution latently. The predatory Morlocks symbolize the inherent change and conflict of life. Despite the future state's "mechanical perfection," the Time Traveller acknowledges that it could never have achieved "absolute permanency": "Mother Necessity, who had been staved off for a few thousand years, came back again, and she began below" (2005 [1895], 57). This inevitable breakdown acts as the last condemnation of a society built on *security* and *permanency*. Even if the Eloi had represented humanity unified and untroubled, peaceful mechanisms will pass away with the ephemeral peace.

The Time Machine stretches Wells's theory of the world beyond the limits of plausibility. Some thousand generations have made humanity less flexible and cooperative than chimpanzees. Before that happened, we supposedly conquered the ecosystem to the extent that "diseases had been stamped out" and "the processes of putrefaction and decay had been profoundly affected" (23). Life's inherent change and conflict seem forgotten in the face of the Eloi's long rule. Wells's focus on the Victorian class system made him reduce all social conflict to class struggle, so that readers are asked to believe that society became peaceful when the rulers and workers were stabilized in isolation from each other (57). The Time Traveller's conversations with his dinner company suggest a greater understanding of social behavior than the simplistic class struggle premise of his story. In the argumentative red-haired Filby, the spasmodic "very young man" and the ponderous "Provincial Mayor," Wells intuitively depicted the within-group variations of age, personality, and intelligence—and the necessarily conflicting interests at every level of interaction—that could have corrected his social theory had they been translated into its language (3, 4). In short, Wells made cardboard copies of evolution and human society to get a stage for his social critique. But the cardboard copies were realistic enough to elicit responses, and the story he set on the stage was ingeniously original.

For readers attentive to imaginative culture, there is an obvious contradiction woven into the fabric of *The Time Machine*. The novel reduces the arts to the decadence of the Eloi. Yet, it approaches the human tragedy with significant stylistic verve, and it is resonant with phrases and images from Tennyson. As the import of the Morlocks is revealed, the Time Traveller bitterly evokes the "heir of all the ages, in the foremost files of time": "this bleached, obscene, nocturnal Thing ... was also heir to all the ages" (Tennyson 1994 [1842], 154; Wells 2005 [1895], 35). In the "eternal sunset" where the Time Traveller watches "the sun grow larger and duller in the westward sky, and the life of the old earth ebb away," there is something of Tennyson's "murmurs of the dying sun"—his mythological images of science's sinking "nebulous star" that grew "broader towards his death and fell" (Wells 2005 [1895], 61; Tennyson 1867, 287; 1994b, 237, 238; for Tennyson's mythological images inspired by science, see Gold 2012, 44–53). And during the Time Traveller's first dash through time, the landscape is "misty and vague" with "trees growing and changing like puffs of vapour," the earth seems "changed—melting and flowing under [his] eyes":

> The hills are shadows, and they flow
> From form to form, and nothing stands;
> They melt like mist, the solid lands,
> Like clouds they shape themselves and go.
> (Tennyson 1994a, 357)

Wells uses art to examine the riddle of man without making art a factor in the riddle. The Time Traveller exemplifies this attitude when he faces the Eloi's abandoned library, dismissing literature with a lyrical flourish: "Had I been a literary man I might, perhaps, have moralized upon the futility of all ambition. But as it was, the thing that struck me with keenest force was the enormous waste of labour to which this somber wilderness of rotting paper testified" (2005 [1895], 49). This self-contradictory attitude to imaginative culture, integral to Wells's worldview, adds another layer to the novel's implausibility.

Ultimately, *The Time Machine* does not require its readers to believe the details of its scientific propositions. The Time Traveller advises his listeners to avoid argument and take his account "as a lie—or a prophecy," a dream or a fiction hatched after "speculating upon the destinies of our race" (63). In tune with Wells, he finishes on a challenge: "And taking it as a story, what do you think of it?" (63). The Time Traveller's dinner guests do not rise to this challenge. All the representatives of Victorian respectability—the Medical Man, the Editor, the Psychologist, the Mayor—remain non-committal or dismissive. Only the nondescript narrator, whose chief characteristic is that he offers the Time Traveller sympathetic attention, gets a say. He makes the future bearable by deciding that the Eloi and Morlocks are not "man's culminating time" (66). He switches focus to the "manhood of the race" that must have occurred before its decline (66). He buffers his belief by declaring that the Time Traveller is a pessimist—someone who always saw civilization as "a foolish heaping that must inevitably fall back upon and destroy its maker" (66). This claim is contradicted by the Time Traveller's own account of his expectations and disappointments when he meets the Eloi: "I had always anticipated that the people of the year Eight Hundred and Two Thousand odd would be incredibly in front of us in knowledge, art, everything" (18). Even if it were true, however, it could only be comforting if the Time Traveller had lied or fundamentally misrepresented his experience. The narrator never suggests any such doubts. Instead, he clings to two flowers that the Time Traveller claims to have gotten from the Eloi girl Weena: "strange white flowers ... to witness that even when mind and strength had gone, gratitude and mutual tenderness still lived on in the heart of man" (66).

The narrator's solitary note of optimism does little to cancel the novel's effect. Two dried flowers and the faith of a nonentity compete with a phantasmagoria of nightmares proceeding from short-sighted human societies and the relentless grind of natural selection. There are the two alien species scuttling and shuffling in "the sunset of mankind" (23). There is the museum, overrun with their dusty footsteps, slowly reclaimed by the "inevitable process of decay" (47). There is the beach in the far future on which the Time Traveller witnesses the last life twitching fitfully, tentacles trailing, "black against the weltering blood-red water" (62). And there is the fact that the Time Traveller does not return from his second journey. Just before the narrator affirms his faith in humanity, he too evokes the perspective of geological time:

> One cannot choose but wonder. Will he ever return? It may be that he swept back into the past, and fell among the blood-drinking, hairy savages of the Age of Unpolished Stone; into the abysses of the Cretaceous Sea; or among the grotesque saurians, the huge reptilian brutes of the Jurassic times. He may even now—if I may use the phrase—be wandering on some plesiosaurus-haunted Oolitic coral reef, or beside the lonely saline seas of the Triassic Age. Or did he go forward, into one of the nearer ages, in which men are still men, but with the riddles of our own time answered and its wearisome problems solved? (66)

In this perspective, humanity seems very small. It seems to flicker and go, despite its brief glory. It is not securely meaningful or uplifting, despite the affections of an anomalous Eloi for a doomed time traveller.

The Eloi and Morlocks followed Wells throughout his life. They represent opposite poles of his fear for humanity: feeble domestication and brute savagery. In different emotional registers, they appear as the delicate drivers of the "Land Ironclads" (1903), as the monstrous Neanderthals of "The Grisly Folk" (1921), or as a deranged strategy for humanity to survive the Martian attack in *War of the Worlds*: "The tame ones will go like all tame beasts; in a few generations they'll be big, beautiful, rich-blooded, stupid—rubbish! The risk is that we who keep wild will go savage—degenerate into a sort of big, savage rat. … You see, how I mean to live is underground" (1978, 371). Eloi and Morlocks together represent one aspect of his dualistic worldview: humanity as "little animals," robbed of dignity, left to "snatch each little scrap of happiness and live and suffer and pass, mattering no more than all the other animals do or have done" (1936).

As Wells translated his theoretical notions into fiction, he engaged his intuitive psychology and made his readers dream a good gripping dream about our species in the abyss of time. He registered the Morlockian dangers that mutate in the darkness of our environment. He evoked the fragility of the human adaptive niche—the cumulative culture that is capable of conquering nature, but could so easily be reduced to the untended "vestiges of books" (49). His political gospel had not yet crystallized into a hardness that pushed the fate of man into easy solutions. The total imaginative effect of *The Time Machine*—as of so much of Wells's early fiction—is a combination of immersive realism, a slightly blurred, dizzying cosmic vision, and mischievously tugging undercurrents from the World State. His first novel was followed by *The Wonderful Visit*, which departs from science fiction to depict an angel passing judgment on humanity. But when Wells returned to evolutionary myth-making, in his third novel, he produced one of his most disturbing myths: *The Island of Doctor Moreau*.

6.4 *The Island of Doctor Moreau* (1896)

Wells claimed that *The Island of Doctor Moreau* was his best attempt to express a "vision of the aimless torture of creation," the "grimace" in which the universe occasionally projected itself at him (1978 [1934], v). The novel is a powerful evocation of his pessimistic mode. It tells the story of Prendick, an everyman with a smattering of biological education, who is shipwrecked, rescued by a passing ship, and brought to a remote island. On this island, the notorious Doctor Moreau—exiled from London after unscrupulous experiments—lives with his assistant Montgomery and a number of misshapen servants. Prendick is disturbed and puzzled by the servants. He soon discovers that they are the result of Moreau's experiments with vivisection: animals that have been imperfectly transformed into humans. Prendick listens in horror to the doctor's explanations, then watches the island's delicate balance break down. The vivisected "Beast People" struggle with the laws imposed by Moreau and his assistant (1978 [1896], 102). When Moreau is killed by his latest experiment, his assistant drunkenly burns their boats and dies in a brawl with the Beast People. Prendick is left alone to witness the final dissolution of the doctor's experiment. He sees the Beast People degenerate, discard their humanity, and return to their predatory ways. When he finally escapes to civilization, he sees so much of the Beast Folk in his fellows that he has to isolate himself from the rest of humanity.

Doctor Moreau himself is at the center of the novel. He is "as remorseless as Nature," embodying both evolution's aimless exploration of "the plasticity of living shapes" and the cruel Creator implied in the grimace that Wells saw (1978 [1896], 113, 116). But there is also a trace of Wells himself—the scientific curiosity, the frustrated ambition to transform animals into rational creatures. The doctor declares that his only goal is to explore the "extreme limit of plasticity in a living shape" (116). However, through an impulse that Prendick considers "a strange wickedness," Moreau has chosen—out of all the possibilities—to transform the beasts into humans (114). According to Moreau, the choice was made by "chance" (114). But in his guise as nonchalant god and Wellsian utopian engineer, he also supposes "that there is something in the human form that appeals to the artistic turn of mind more powerfully than any animal shape can" (114). He is at once playing and pursuing purpose, exploring and perfecting. And in all of his modes, he is ruthless.

Allegorically, the island functions as "the whole balance of human life in miniature, the whole interplay of instinct, reason, and fate, in its simplest form" (131). Whether as a creator or as the aimless force of nature, Moreau is the tormentor of living things—particularly because he creates human life. Through a combination of his indefinite artistic calling and a puzzle-solving compulsion, he strives to conquer "the stubborn beast flesh," to "burn out all the animal" and to "make a rational creature of [his] own" (117, 118). Thus, as Prendick puts it, he takes animals with "instincts fitly adapted to their surroundings, as happy as living things may be" and makes them "stumble in the shackles of humanity" living "in a fear that never die[s], fretted by a law they [cannot] understand" (131). They endure an existence that begins in agony and is "one long internal struggle" until they die (131). That is the dream of the human intellect in Wells's pessimistic mode—the "ambition" and "labour" for which the Time Traveller mourned (2005 [1895], 49). The microcosm of Moreau's island tacitly identifies the mechanical Eloi and Morlocks as the lucky ones.

In Prendick's voice, Wells curses the wantonness, vile cruelty, aimlessness, and waste of the spectacle of human life. But Wells's voice is also closely aligned with that of Moreau. A year before the novel was published, Wells described the theoretical rationale behind Moreau's vivisectionist project in a scientific article called "The Limits of Individual Plasticity" (1975 [1895]b). Another year back, Wells published an article called "The Province of Pain," which expressed Moreau's teleological idea of pain as "a phase through which life must pass on its evolution from the automatic to the spiritual" (1975 [1894], 199). Like Moreau, Wells saw himself as a "student of Nature," with an imagination that ran wild on the gruesome possibilities of the natural world (1978 [1896], 116). At times, he pretended to something of the same ruthlessly detached perspective, ignoring social conventions and upsetting those around him by asking "Why not?" (1934, 392–405). Like Moreau, Wells had an obsessive desire to turn "little animals" into rational creatures.

Prendick's human sympathy might be said to represent the darker part of Wells's pessimism. Where Moreau sees a quasi-teleological vision of spiritualizing evolution that he can influence, Prendick sees suffering creatures that would have been better off as beasts unconsciously guided by instinct. It is Moreau—"a religious man... as every sane man must be"—who holds onto Wells's dualistic worldview through most of the novel

(115). He berates Prendick as a "materialist" and as an animal who has failed to detach himself from instinctive sympathy, thus merely "thinking a little less obscurely what an animal feels" (115). The logical implication is straightforward: as long as thinking is influenced by biologically based emotion, it is the thinking of an animal rather than of a human being—of "natural" rather than "artificial" man. That is an extreme form of Wells's dualism. But Prendick, too, protests "hotly" against being characterized as a materialist (115). It is with Prendick's voice that Wells ends the story, seeking something "more than animal" in the stars (156).

Prendick's and Moreau's perspectives fight for Wells's voice throughout the novel. Moreau's oscillation between dualistic teleology and curiosity about aimless adaptive plasticity reflects Wells's own ambivalence. Prendick spends much of his time coming to terms with Moreau's vision of the world—failing to grasp the spiritual ideal of rationality, horrified by the conscious cruelty, and disturbed also, conversely, by the "mad, aimless investigations" and the "wantonness" of the suffering (131). Moreau has the clarity of mind that Wells reserved for scientific articles, the unflinching, "remorseless" perspective on the natural world (116). But his remorselessness limits the width of his perspective even as it increases his penetration. He sees the oddness and complexity of all biological life, but he does not see the springs of human morality: cooperative coalitions, prosocial self-sacrifice, mammalian mother-infant affection, affiliative bonding, kin protection and pair bonding. Thus, he relegates any impulse that is not base, egocentric, irrational and violent to a "soul" set apart from the natural world (119). Prendick has the sympathetic vision—the freer access to Wells's intuitive insight into other minds, and consequently something approximating wisdom in his psychological observations. In his own laconic phrase: "I have an imagination." (78). He sees the reflection of humanity in the beast people, and reacts with fear, pity, frustration, sorrow, and anger. When he gradually comes to understand life on the island, he sees the full horror of the human condition as it appeared to Wells—"now and then," though he "rarely" admitted it—when the universe projected itself in a grimace (1978 [1934], v).

Culture on the island is primarily represented by Moreau's rational ideal. The beast people—all let loose to fend for themselves as they fail to satisfy Moreau's standard for humanity—form a small society that is "a kind of travesty of humanity" (118). There, they lead a "mockery of a rational life," guided by a set of prohibitions that they call "the Law" (119). Though the Law was not put in place by Moreau himself, but by one of his servants ("one among the boys, a bit of a missionary" 117), it follows Moreau's general ideal. It forbids the beast people to indulge their animal instincts—to claw at trees, to walk on all fours, to chase each other, to eat flesh—and it ends each commandment with the ominous question, "Are we not Men?" (105). Moreau enforces these commandments by punishing transgressors, and the beast people live in fear of his "House of Pain" (105).

The beast people's Law is an overt parallel to religious doctrines, but it also parallels imposed social conventions generally. It is the conventionality against which Wells raised his "Why not?" in novels like the free love manifesto *Ann Veronica* (the female beast folk were "liable to much furtive persecution in spite of the monogamy the Law enjoined" 121). The Law also represents "morality," the "padding of suggested emotions and habits" that Wells judged necessary for civilizing "Paeleolithic man" (1975 [1897], 221). The pessimistic vision that Moreau's island offers of civilization is nothing like the tragedy of *The Time Machine*, or like Wells's later paean to the Mind of the Race. It most closely resembles the "blind submission" and "sheer brutality of the accepted thing" that haunts the protagonist's working-class mother in *Days of the Comet:*

> She seemed always to be wanting to say to me, "Dear, I know it's hard—but revolt is harder. Don't make war on it, dear—don't! Don't do anything to offend it. I'm sure it will hurt you if you do—it will hurt you if you do." (1978 [1900], 712)

Moreau expresses Wells's scorn for planless natural man more directly and venomously than most of Wells's characters. He has pursued a rationality that transcends pain, and accidentally created a travesty of human society that is upheld entirely by pain. But as he looks at the beast people, he sees "nothing but the souls of beasts, beasts that perish—anger, and the lusts to live and gratify themselves" (119). The same disgust is often applied to humanity in Wells's work, from *Anticipations* to the speech against "little animals" in *Things to Come*. Four years after *Doctor Moreau*, the narrator of *Days of the Comet* described society as "an aimless thing that ate and drank and slept and bore arms, and was inordinately proud of itself because it had chanced to happen" (1978 [1900], 820). For Moreau—as implicitly for Prendick, at the end—these short-term pursuits make the beast people less than human.

Moreau's frustration with his experiments is, in one sense, unreasonable. There is no qualitative difference between human beings and other animals—no distinction like the "souls" for which his scalpel is digging (119). Much less is there a qualitative difference between human beings and the beast people. The Ape man, for instance, is a clearly recognizable human type:

> He had an idea, I believe, that to gabble about names that meant nothing was the proper use of speech. He called it "Big Thinks" to distinguish it from "Little Thinks," the sane every-day interests of life. If ever I made a remark he did not understand, he would praise it very much, ask me to say it again, learn it by heart, and go off repeating it, with a word wrong here or there, to all the milder of the Beast People. He thought nothing of what was plain and comprehensible. (149, 150)

Nonetheless, Moreau's sense that the beast people are inadequate—like Prendick's desperate rejection of their resemblance to humanity at the end of the novel—has some justification. There are distinct differences between the beast people and real human beings.

Though the satire in *Moreau* is sometimes brilliant, Wells painted the human condition in its "simplest form," using his interpretation of evolution (131). In line with Huxleyan ethics, the beast people's natural dispositions seem to be primarily violent and egocentric. Though the beast people marry and procreate, we never see them display affection toward one another, form families or long-term cooperative coalitions, protest when Moreau takes their offspring to the House of Pain, or mourn when the offspring return vivisected. Following Wells's idea that civilization would disappear overnight if the immaterial "mental factor" were destroyed, the beast people degenerate rapidly once the Law is removed (1975 [1897], 221). When they are freed from Moreau's punishing hand, and bereft of the beast creature that repeated the Law to them, they simply disperse into solitary habitation. There seem to be no impulses toward forming smaller tribes or creating new laws. Finally, they have no culture that appeals to their positive emotions, channels constructive instincts, and allows them to develop their personal identities. The Law represents imposed morality, religion, and social conventions—but it does not adequately represent imaginative culture. The beast people lack the ability to create their own fictional worlds to help them live their lives.

The Law has an emotional aspect, if not an imaginative one. It has traces of a primitive mythology similar to that of "A Story of the Stone Age": the beast people turn Moreau, his assistant Montgomery, Prendick, and Moreau's laboratory into symbols. Their ceremonial recital of the Law has the quality of religious ecstasy—a Dionysian abandon as they are "swaying in unison and chanting" (105). But the commandments and symbols are either externally imposed or crudely descriptive ("The Master", "the Other with the whip," "The Man with the Bandaged Arm" 124, 148). The mythology of the Law does not lead the beast people to new ideas or new interpretations of the world. Its main function is prohibitive, and its emotional register is exalted fear. Unlike the "Stone Age" people's mythology—but very like the "accepted" religion of the mother from *Days of the Comet*—the Law keeps the beast people precariously suspended between base instinct and terror-induced discipline (1978 [1900], 712). That suspension was one of Wells's most consistent ideas about the human condition. In *Doctor Moreau*, he played it out in full force. As a result, he pictured a world in which imaginative culture is at best incidental. His own attraction to imaginative culture—his immersion in it and creation of it—worked against his theoretical view to give the novel a profoundly self-conflicted air: an air not of articulated though unresolved conflict, but of intellectual and artistic incompleteness. He did not, to use W. H. Auden's words, achieve the artistic ideal that is "a clear expression of mixed feelings" (1946).

Imaginative culture was negligible in Wells's "simplest" portrait of the human condition. It was not a basic force in the interplay, like "instinct," "reason" or "fate" (131). Yet, *Doctor Moreau* is itself a therapeutic fictional world. It allowed Wells to face his nightmare version of the natural world—to give it shape, to make his way through it vicariously, and to conclude on an elevating view of the stars. The novel is adorned with telltale signs of his own literary immersion. Its ending mirrors the ending of *Gulliver's Travels*. When Prendick comes back to civilization after his experience on the island, he shrinks in terror and disgust from human beings, seeing in them "another, still passably human, Beast People" with "faces keen and bright... dull or dangerous... unsteady, insincere" and liable to "degradation" (Wells 1978 [1896], 155). When Gulliver returns from the Country of the Houyhnhnms, where rational horse creatures rule over base primate creatures called Yahoos, he fears the "teeth" and "claws" of "English Yahoos," and has to force himself to look in a mirror in order to "habituate [himself] by time to tolerate the sight of a human creature" (Swift 2011 [1726], 373, 375). Even as Moreau dips his specimens "into the bath of burning pain" to "burn out all the animal," Tennyson's poetry reverberates in the background: "hot with burning fears... in baths of hissing tears," moving "upward, working out the beast" (Wells 1978 [1896], 118; Tennyson 1867, 183). In contrast to the imaginatively vacuous beast people, Wells had an inventive inner life pervaded by the imaginative constructs of his predecessors and contemporaries.

On his return to civilization, Prendick tries to convince himself that the people around him are "altogether different from the Beast Folk": "perfectly reasonable creatures, full of human desires and tender solicitude, emancipated from instinct, and the slaves of no fantastic Law" (156). That effort amounts to a tragicomic naivety; as naïf, he is the straight man for Wells's satire of the human condition. But he is also a personification of Wells's own failure of nerve before the grimace projected by the universe. For Prendick, the parallel between the beast people and humanity is unendurable. Wells mocks him because he flinches where Moreau would not have flinched. But the parallel is also unendurable to Wells himself. He both feared the truth of that parallel and intuited, dimly, inconsistently, that there is something radically inadequate about it—inadequate not only as a model for ethics, but as a symbolic figuration of the human condition.

Wells had a fuller intuitive understanding of cooperation and imaginative culture than is contained in the image of the beast people. He gave the "Stone Age" people dreams, symbols, and storytelling dances that crystallized their ideas. He also gave them the thrill of discovery, heroism, love, and companionship as well as warfare, deceit, pettiness, and violent death. He had Moreau's assistant Montgomery save Prendick's life and stocked Moreau's shelves with "editions of the Latin and Greek classics" next to the surgical works (86). He made Prendick attempt, and fail, to read "a crib of Horace" in order to tolerate the cries of the vivisectionist's victim (90). He even gave Moreau an "artistic turn of mind" and made him submit to "a gleam of classical memory" by creating a living "Satyr" (114, 124). And at the end of the novel, after tacitly chiding Prendick for dismissing the reality of animal baseness in human life, he had him look to "the glittering hosts of heaven" for the missing piece of the puzzle.

Prendick looks for the greatness of man in something greater than man. He mistakes transcending the self for transcending the species, and finds his means to go beyond "purely personal particulars" in the same place that Boon found the Mind of the Race (1915, 171). Thus, he is more or less implicated in Wells's dualistic perspective. He dismisses the beast people as inadequate, but he does not do it in quite the same way as Moreau did. When Moreau glimpsed a "kind of upward striving" in the beast people, "part vanity, part waste sexual emotion, part waste curiosity," he measured it against his spiritual ideal of rationality and dismissed it as mockery (1978 [1896], 119). Moreau only wanted a "rational creature," free of its "beast flesh" (117, 118). Thus, he kept reaching in vain for something vague that needed altering "in the seat of the emotions" (118). It had to do with "desires that harm humanity," "anger, hate, or fear" (118). In order to reach it, he might have needed a supernatural intervention—like the extraterrestrial gas from *Days of the Comet* that brought about the World State by robbing humanity of "its spites, its intense jealousies, its inconsistencies, its humour" (1978 [1900], 857).

With his spiritual ideal in mind, Moreau obsessively dipped living creatures into the "bath of burning pain," wishing they would wriggle out of the "dust" and ascend into the realm of reason (116, 118). That realm was occupied by the Martians of *War of the Worlds*, the Selenites of *First Men in the Moon*, and the dehumanized humanity of Wells's various utopias. Prendick, in contrast, needs more. He does not content himself with stargazing, but spends his days "surrounded by wise books,—bright windows in this life of ours, lit by the shining souls of men" (156). In that regard, he was closer to the wisdom Wells displayed in practice than to the frustrated inconsistencies he displayed in theory.

6.5 Conclusion

Wells's evolutionary myths remain his most enduring cultural contribution. Though his early speculative fiction includes some memorable non-evolutionary stories, like *The Invisible Man* and "The Country of the Blind," those stories are not as widespread and appreciated as *The Time Machine*, *The Island of Doctor Moreau*, and *The War of the Worlds*. Those three novels have a living history of critical accolades, scholarly discussion, and adaptations into other media (Aldiss 1986; Collinge 2017; Glendening 2002; Hovanec 2015; Quade 2007). It was *The Island of Doctor Moreau* that became the cult movie *Island of Lost Souls* in 1932, and *The War of the Worlds* that Orson Welles broadcast in 1938 to terrify his countrymen. This audience response is out of sync with Wells's own preference for his social realist fiction, but it is in sync with his focus on evolution.

Wells never stopped pursuing the ideal represented by Huxley's biology class: "a vivid, sustained attempt to see life clearly and to see it whole, to see into it, to see its inter-connexions, to find out, so far as terms were available, what it was, where it came from, what it was doing and where it was going" (Wells 1934, 169). This scientific quest remained at the core of his career as a literary author. Throughout his massive *Experiment in Autobiography,* he mentions *biology* 47 times, *evolution* 133 times, *science* 199 times, and *literature* only 32 times. The Huxleyan quest resounds through his books, from his first *Textbook of Biology* onward, and even surfaces in social realist novels like *Love and Mr Lewisham:* "Natural Selection—it follows ... this way is happiness ... must be" (1899, 322). But it was nowhere given as complete an expression as in *The Time Machine* and *The Island of Doctor Moreau.*

Wells is often given his due either grudgingly or apologetically. Henry James advanced under his "spell" in spite of himself (quoted in Wells 1934, 413). George Orwell pilloried the World State but granted that "The minds of all of us, and therefore the physical world, would be perceptibly different if Wells had never existed" (1941). Harold Bloom dismissed the "failed prophet," "solipsistic and libido-driven," but admitted that Wells had dreamed "inescapable nightmares" (2005, xii, xiii). John Huntington claimed that Wells was "greater than he himself understood" because of his "deep conflict between ideal rationality and a much less admired, though not therefore contemptible, emotionality" (2005, 68). Even Theodore Roosevelt expressed a fascination closely hounded by frustration:

> He mentioned my *Time Machine*... He became gesticulatory, and his straining voice a note higher in denying the pessimism of that book as a credible interpretation of destiny ... "Suppose, after all," he said slowly, "that should prove to be right, and it all ends in your butterflies and morlocks. *That doesn't matter now.* The effort's real. It's worth going on with. It's worth it—even so." (Wells 1934, 648, 649)

Wells's limitations were real, and his assertive attitude may have exacerbated negative responses. Orwell added a humbling barb to his accolade: "How much influence any mere writer has, especially a 'popular' writer ... is questionable" (1941). Virginia Woolf famously placed Wells together with John Galsworthy and Arnold Bennet below the authors of "classics," in a separate category of "everyday imperfection": authors who have "shown us what they might have done but have not done; what we certainly could not do, but as certainly, perhaps, do not wish to do" (1984 [1925], 147). Nevertheless, the co-father of science fiction has consistently attracted enthusiastic admirers, from William James to his own illegitimate son (Bell 2018b; West 1984). Warran Wagar expresses Wells's relation to dystopian fiction with exaggeration but not pure absurdity: "Nothing foreseen by the Zamyatins, Huxleys, and Orwells of late generations was missed by Wells in the science fiction and studies of the human future that he published before 1914" (1989, 393).

Wells's stories were shaped by his ideas about human nature. The stories also helped shape his worldview, which directed his own behavior. The intentions and effects of his stories become clearer if we understand how his self-narrative helped him face the psychological challenge of evolution. The Eloi and Morlocks are not just general representations of the fragility of the human species; they are psychologically vital symbols of choosing the lowly natural impulses that would lead to destruction. *Moreau* is not only the nightmare of a post-Darwinian world, but of Wells's approach to that Darwinian world: his personal nightmare about the futility of his pursuit of the World State. The strong but frustrating effect of such symbolic visions makes sense in light of the strengths and limitations of Wells's worldview. We might be carried along by the intuitive excitement of Wells's dualistic Darwinian myth, charmed by its historical peculiarities, and impressed by the elements of symbolic truth in his fiction, but still find ourselves reacting against the incoherence that results from his broad-stroked split view of humanity.

The World State against which Wells measured reality was an impossible ideal. It was premised on the hypothesis that humans can disconnect from their evolved social dynamics and basic motivations. That hypothesis was repeatedly refuted, for one thing, by Wells's attempts to experience free love (Wells 2008). As he put it, he wanted "the primary desires and emotions, the drama of the immediate individual life" to become "subordinate more and more, generation by generation, to beauty and truth, to universal interests and mightier aims" (1934, 7). But in expressing this desire, he was himself fuelled by primary desires and emotions: his need to see himself as the protagonist in his own life story, his need to partake in a constructive egalitarian community, and his need to understand the forces and values of the world, making them meaningful through vivid imagery

(McAdams 2019; Boehm 1997; Carroll 2012; Carroll et al. 2012). He was not unaware of his internal contradictions. In *World Brain* (1938), he acknowledges the shortcomings of his ideas about the world—proceeding immediately, however, to withdraw the apology:

> All these attempts had profound and conspicuous faults and weaknesses; even my friends are apt to mention them with an apologetic smile; presumptuous and preposterous they were, I admit, but I look back at them, completely unabashed. Somebody had to break the ice. Somebody had to try out such summaries on the general mind. My reply to the superior critic has always been—forgive me—"Damn you, do it better." (4)

Thus speaks Wells the shameless humorist, and Wells the man of faith. The shameless humor often supported the faith, but was—in any case—never far away. In the witty visionary passages of *Boon*, Wells celebrates the self-mocking drawings produced by his avatar Boon: "They were, I know, as necessary to his whole conception as his swinging reference to the stars; both were as much part of his profound belief as the gargoyle on the spire and the high altar are necessary parts of a Gothic cathedral" (1915, 175). Toward the end of this passage, Wells presses Boon in Socratic dialogue until the only argument against the non-existence of the Mind of the Race is "Because I will not have it so" (198, 199).

Wells's success depended on his acute intuitive understanding of the human imagination. Sometimes, he did turn that intuition against his tendency to stick (as he put it) "close to the trail that leads to the World State" (2008, 102). The epilogue of *Days of the Comet* acknowledges the imaginative limitations of his utopian vision with remarkable clarity. A comet has burned away human emotions and produced a new world order. The ordinary human narrator is struggling to understand his guide to this new world:

> At the end ... I felt his story had slipped away from my sympathies altogether. Those Beltane fires had burnt something in him that worked living still and unsubdued in me, that rebelled ... I became inattentive. I no longer felt with him, nor gathered a sense of complete understanding from his phrases. ... He and these transfigured people—they were beautiful and noble people, like the people one sees in great pictures, like the gods of noble sculpture, but they had no nearer fellowship than these to competitive men. As the Change was realised, with every stage of realisation the

gulf widened and it was harder to follow his word. (1978 [1900], 857, 858)

Perhaps this was a reaction to the response of contemporary readers—an attempt to pre-empt and redirect their response. Perhaps it was a description of his own experience when he tried to imagine a full cycle of the supernormal, transfigured life. Plausibly, it was both.

Despite his self-awareness, Wells did not change the main story he told himself about humanity (2008). He simply alternated between the role of a tragic hero and a triumphant one. If his utopian visions could not appeal to the imagination of competitive men, it was the men needed changing—and the change, if it could be effected, could be effected by other means than imagination. It would come with the rational progression of the Mind of the Race, through which "sanity grows" (Wells 1915, 198). He reassured himself against the limitations of his perspective by persistently splitting reason and emotion, underestimating the power of imaginative motivation. Nevertheless, when he was at his most disillusioned, it was to the imaginative value of the utopian vision that he appealed—claiming that it was, at least, "a helpful dream" (1934, 8). The "transfigured people" of the World State may not have had much imaginative appeal even for Wells, but his life was directed by the eternally ordered stars that represented their self-transcendence, and by the aimless animalistic needs and cares from which they represented an escape.

Wells's worldview and early fiction offer a peculiar lesson about imagination. The spiritualized political theory that Wells himself considered the greatest achievement of his life has been largely forgotten or repudiated, overshadowed by the "playful parables" he wrote to get on as a young man. Those playful parables were replete with vivid symbolic images that helped make the forces of a changing world comprehensible to him and his readers. The parables were shaped by keen psychological intuition and bold imagination, but also by a worldview that did not recognize the power of intuition and imagination. They are a complex mixture of unflinching curiosity and single-minded escapism, symbolic truth, historical misconceptions, prophecy, delusion, and wilful distortion. Taken in all, they are a lesson about the double-edged nature of imaginative culture. Wells's good gripping dreams carried him in multiple directions. The dualistic vision may have afforded him comfort by doggedly training his eyes on the stars, but it also blinded him to basic forces at work in his own life—forces that, had he channeled them, might have helped him achieve some of his desired political change. When he was giving his mind freer play, the stories he told revealed more of the world to him than his theorizing. But like Prendick, he had an inadequate vision of the human species and needed to survive by looking for hope in the stars. With or without Wells's main story in mind, readers may experience that same double vision. In those guiding and misguiding stars of Wells, we may see our own useful dreams, nightmares, outlines of the world, and satirical self-portraits. In that sense, as far as there is a Mind of the Race, Wells is part of it.

References

Aldiss, Brian. 1986. "The Great General in Dreamland: H. G. Wells." In *Trillion Year Spree,* by Brian Aldiss and David Wingrove, 145–66. New York: Atheneum.

Alkon, Paul K. 2016. *Science Fiction Before 1900: Imagination Discovers Technology*. London: Routledge.

Asimov, Isaac. 1968. "The Contribution of H. G. Wells." In *H. G. Wells* The Time Machine/The War of the Worlds, 9–22. Greenwich: Fawcet.

Auden, W. H. 1946. *New Year Letter.* London: Faber and Faber.

Barnett, Richard. 2006. "Education or Degeneration: E. Ray Lankester, H. G. Wells and *The Outline of History.*" *Studies in History and Philosophy of Biological and Biomedical Sciences* 37: 203–29. https://doi.org/10.1016/j.shpsc.2006.03.002.

Beenstock, Zoe. 2015. "Empiricist Political Theory and the Modern Novel: The Social Contract and H. G. Wells." *Modern Language Quarterly* 76 (1): 57–77. https://doi.org/10.1215/00267929-2827549.

Bell, Duncan. 2018a. "Founding the World State: H. G. Wells on Empire and the English-Speaking Peoples." *International Studies Quarterly* 62 (4): 867–79. https://doi.org/10.1093/isq/sqy041.

———. 2018b. "Pragmatism and Prophecy: H. G. Wells and the Metaphysics of Socialism." *American Political Science Review* 112 (2): 409–22. https://doi.org/10.1017/S0003055417000508.

Bjorklund, David F. 1997. "The Role of Immaturity in Human Development." *Psychological Bulletin* 122 (2): 153–69. https://doi.org/10.1037/0033-2909.122.2.153.

Bloom, Harold. 2005. "Introduction." In *Bloom's Modern Critical Views: H. G. Wells,* edited by Harold Bloom, ix–xii. New York: Chelsea House Publishers.

Boehm, Christopher. 1997. "Impact of the Human Egalitarian Syndrome on Darwinian Selection Mechanics." *The American Naturalist* 150 (S1): 100–21. https://doi.org/10.1086/286052.

Carroll, Joseph. 2012. "The Adaptive Function of the Arts: Alternative Evolutionary Hypotheses." In *Telling Stories: Literature and Evolution*, edited by Gansel, Carsten and Dirk Vanderbeke, 50–63. Berlin: De Gruyter.

Carroll, Joseph, Jonathan Gottschall, John Johnson, and Daniel J. Kruger. 2012. *Graphing Jane Austen: The Evolutionary Basis of Literary Meaning*. Cognitive Studies in Literature and Performance. New York: Palgrave Macmillan.

Collinge, James T. 2017. "'With Envious Eyes': Rabbit-Poaching and Class Conflict in H. G. Wells's *The Time Machine* and *The Island of Doctor Moreau*." *Literature & History* 26 (1): 39–55. https://doi.org/10.1177/0306197317695082.

Delbanco, Nicholas. 1990. *Group Portrait: Joseph Conrad, Stephen Crane, Ford Madox Ford, Henry James, and H. G. Wells*. New York: Carroll & Graf.

Di Leo, Jeffrey R. 2017. "Catastrophic Education: Saving the World with H. G. Wells." *The Comparatist* 41: 153–76. https://doi.org/10.1353/com.2017.0009.

Earle, Edward Mead. 1950. "H. G. Wells, British Patriot in Search of a World State." *World Politics* 2 (2): 181–208. https://doi.org/10.2307/2009188.

Forster, E. M. 2011 [1909]. *The Machine Stops*. London: Penguin.

Glendening, John. 2002. "'Green Confusion': Evolution and Entanglement in H. G. Wells's *The Island of Doctor Moreau*." *Victorian Literature and Culture* 30 (2): 571–97. https://doi.org/10.1017/S1060150302302109h.

Gold, Barri J. 2012. *ThermoPoetics: Energy in Victorian Literature and Science*. Cambridge: MIT Press.

Haley, James L. 2010. *Wolf: The Lives of Jack London*. New York: Basic Books.

Henrich, Joseph. 2015. *The Secret of Our Success: How Culture is Driving Human Evolution, Domesticating our Species, and Making Us Smarter*. Princeton: Princeton University Press.

Higgins, Richard. 2008. "Feeling Like a Clerk in H. G. Wells." *Victorian Studies* 50 (3): 457–75. https://doi.org/10.2979/VIC.2008.50.3.457.

Hill, Kim, Michael Barton, and A. Magdalena Hurtado. 2009. "The Emergence of Human Uniqueness: Characters Underlying Behavioral Modernity." *Evolutionary Anthropology: Issues, News, and Reviews* 18 (5): 187–200. https://doi.org/10.1002/evan.20224.

Hillegas, Mark R. 1967. *The Future as Nightmare: H. G. Wells and the Anti-Utopians*. New York: Oxford University Press.

Hovanec, Caroline. 2015. "Rereading H. G. Wells's *The Time Machine:* Empiricism, Aestheticism, Modernism." *English Literature in Transition, 1880–1920* 58 (4): 459–85.

Hume, Kathryn. 1990. "Eat or be Eaten: H. G. Wells's 'Time Machine.'" *Philological Quarterly* 69 (2): 233–51.

Huntington, John. 2005. "H. G. Wells: Problems of an Amorous Utopian." In *Bloom's Modern Critical Views: H. G. Wells* edited by Harold Bloom, 67–80. New York: Chelsea House Publishers.
Huxley, Aldous. 2007 [1932]. *Brave New World*. London: Vintage Books.
Huxley, T. H. 2009. [1863]. *Evidence as to Man's Place in Nature*. Cambridge: Cambridge University Press.
———. 2009 [1893]. *Evolution and Ethics: Delivered in the Sheldonian Theatre, May 18, 1893*. Cambridge: Cambridge University Press.
Library of Illinois. "Text-book of Biology." *H. G. Wells, Time Traveler, Teacher to the World*. Accessed on September 8, 2019. https://www.library.illinois.edu/rbx/hgwells2016/2016/09/text-book-of-biology/.
London, Jack. 1982 [1913]. "John Barleycorn." In *Novels and Social Writings*, edited by Donald Pizer, 933–1112. New York: The Library of America.
MacKenzie, 1973. Norman and Jeanne MacKenzie. *H. G. Wells: A Biography*. New York: Touchstone Books.
Maxwell, Anne. 2012. "'Becoming Martianized': Liberal Eugenics and H. G. Wells' *Star-Begotten*." *English Studies* 93 (6): 683–99. https://doi.org/10.1080/0013838X.2012.700574.
McAdams, Dan P. 2019. "'First We Invented Stories, Then They Changed Us': The Evolution of Narrative Identity." *Evolutionary Studies in Imaginative Culture* 3 (1): 1–18. https://doi.org/10.26613/esic.3.1.110.
McLean, Steven. 2009. *The Early Fiction of H. G. Wells: Fantasies of Science*. Basingstoke: Palgrave Macmillan.
———. 2011. "'I Flung Myself into Futurity': Wells Studies for the Twenty-First Century." *Literature Compass* 8 (7): 476–86. https://doi.org/10.1111/j.1741-4113.2011.00818.x.
McNabb, John. 2015. "The Beast Within: H. G. Wells, *The Island of Doctor Moreau*, and Human Evolution in the mid1890s." *Geological Journal* 50: 383–97. https://doi.org/10.1002/gj.2607.
Norenzayan, Aran, Scott Atran, Jason Faulkner, and Mark Schaller. 2006. "Memory and Mystery: The Cultural Selection of Minimally Counterintuitive Narratives." *Cognitive Science* 30 (3): 531–53. https://doi.org/10.1207/s15516709cog0000_68.
Orwell, George. 1941. "Wells, Hitler and the World State." *Horizon* 4 (20): 133–38.
Pudelko, Brygida. 2015. "May Sinclair and H. G. Wells's Involvement in the Suffrage Movement." *Journal of History: Culture and Art Research* 4 (2): 56–70. https://doi.org/10.7596/taksad.v4i2.433.
Quade, Penelope. 2007. "Taming the Beast in the Name of the Father: *The Island of Dr. Moreau* and Wells's Critique of Society's Religious Molding." *Extrapolation* 48 (2): 292–301. https://doi.org/10.3828/extr.2007.48.2.6.

Rabkin, Eric S. 1983. *Science Fiction: A Historical Anthology*. New York: Oxford University Press.

Richerson, Peter J., Robert L. Bettinger, and Robert Boyd. 2008. "Evolution on a Restless Planet: Were Environmental Variability and Environmental Change Major Drivers of Human Evolution?" In *Handbook of Evolution: The Evolution of Living Systems (Including Hominids), Volume 2*, edited by F. M. Wiketits and F. J. Ayala, 233–42. Weinheim: Wiley-VCH Verlag GmbH.

Shepard, Irving M. 1956. "Introduction." In *Jack London's Tales of Adventure*, ed. Irving Shepard. Garden City, NY: Doubleday & Company.

Suvin, Darko. 1977. *H. G. Wells and Modern Science Fiction*. Bucknell University Press.

Swift, Jonathan. 2011 [1726]. *Gulliver's Travels into Several Remote Nations of the World*. Plain Label Books.

Tennyson, Alfred. 1867. *In Memoriam*. London: Edward Moxon & CO.

———. 1994a. "CXXIII." In *The Works of Alfred Lord Tennyson*, 357. London: Wordsworth Editions.

———. 1994b. *The Works of Alfred Lord Tennyson*. London: Wordsworth Editions.

———. 1994 [1842]. "Locksley Hall." In *The Works of Alfred Lord Tennyson*, 149–55. London: Wordsworth Editions.

Thompson, Terry W. 2017. "An Atlantis Allusion in H. G. Wells's *The Country of the Blind*." *The Explicator* 75 (1): 62–64. https://doi.org/10.1080/00144940.2016.1274714.

Upal, M. Afzal. 2011. "Memory, Mystery and Coherence: Does the Presence of 2-3 Counterintuitive Concepts Predict Cultural Success of a Narrative?" *Journal of Cognition and Culture* 11 (1): 23–48. https://doi.org/10.1163/156853711X568671.

Wagar, Warren W. 1989. "H. G. Wells and the Scientific Imagination." *The Virginia Quarterly Review* 65 (3): 390–400.

Wells, H. G. 1975 [1892]. "Ancient Experiment in Cooperation." In *Early Writings in Science and Science Fiction*, edited by Robert M. Philmus and David Y. Hughes, 187–93. Berkeley: University of California Press.

———. 1975 [1894]. "The Province of Pain." In *H. G. Wells: Early Writings in Science and Science Fiction*. Edited by Robert M. Philmus and David Y. Hughes, 194–99. Berkeley: University of California Press.

———. 1975 [1895]-a. "Bio-Optimism." *H. G. Wells: Early Writings in Science and Science Fiction*, edited by Robert M. Philmus and David Y. Hughes, 206–10. Berkeley: University of California Press.

———. 1975 [1895]-b. "The Limits of Individual Plasticity." In *H. G. Wells: Early Writings in Science and Science Fiction*, edited by Robert M. Philmus and David Y. Hughes, 36–39. Berkeley: University of California Press.

———. 1975 [1896]. "Human Evolution, an Artificial Process." *Early Writings in Science and Science Fiction*, edited by Robert M. Philmus and David Y. Hughes, 211–19. Berkeley: University of California Press.

———. 1975 [1897]. "Morals. and Civilisation." In *H. G. Wells: Early Writings in Science and Science Fiction*, edited by Robert M. Philmus and David Y. Hughes, 220–28. Berkeley: University of California Press.

———. 1978 [1896]. "The Island of Dr. Moreau." In *The Complete Science Fiction Treasury of H.G. Wells*, 69–161. New York: Wings Books.

———. 1983 [1897]. "The Star." *Science Fiction: A Historical Anthology*, edited by Eric S. Rabkin. New York: Oxford University Press.

———. 1899. *Love and Mr Lewisham: The Story of a Very Young Couple*. New York: Frederick A. Stokes Company.

———. 1978 [1900]. "In the Days of the Comet." In *The Complete Science Fiction Treasury of H.G. Wells*, 691–860. New York: Wings Books.

———. 1902. *Anticipations of the Reaction of Mechanical and Scientific Progress Upon Human Life and Thought*. 2nd edition. London: Chapman & Hall.

———. 1915. *Boon, the Mind of the Race, the Wild Asses of the Devil, and the Last Trump; Being a First Selection from the Literary Remains of George Boon, Appropriate to the Times, Prepared for Publication by Reginald Bliss. With an Ambiguous Introduction by H. G. Wells*. London: T. F. Unwin.

———. 1917. *God the Invisible King*. New York: Macmillan.

———. 1919. *The Undying Fire*. New York: Macmillan.

———. 1921. *The Outline of History: Being a Plain History of Life and Mankind*. 3rd edition. New York: The MacMillan Company.

———. 2008 [1932]. *The. Bulpington of Blup*. Cornwall: House of Stratus.

———. 1934. *Experiment in Autobiography: Discoveries and Conclusions of a Very Ordinary Brain (Since 1866)*. New York: The Macmillan Company.

———. 1978 [1934]. "Preface" In *The Complete Science Fiction Treasury of H. G. Wells*, iii–3. New York: Wings Books.

———. 1936. *Things to Come*. Screenplay. Directed by William Cameron Menzies. United Artists.

———. 1938. *World Brain*. Garden City: Doubleday.

———. 1946. *Mind at the End of its Tether.* London: Heinermann.

———. 1995 [1942]. *The Conquest of Time*. Amherst: Prometheus Books.

———. 1975. *Selected Short Stories*. London: Penguin Books.

———. 1978. *The Complete Science Fiction Treasury of H.G. Wells*, 691–860. New York: Wings Books.

———. 2005 [1895]. *The Time Machine*, edited by Patrick Parrinder. London: Penguin Books.

———. 2008. *H. G. Wells in Love: Postscript to An Experiment in Autobiography.* Edited by G. P. Wells. London: Faber and Faber.

West, Anthony. 1984. *H. G. Wells: Aspects of a Life*. New York: Random House.

Woolf, Virginia. 1984 [1925]. *The Common Reader.* New York: Harcourt.
Wrangham, Richard. 2019. *The Goodness Paradox: The Strange Relationship between Virtue and Violence in Human Evolution*. New York: Pantheon Books.

CHAPTER 7

Joseph Conrad's Evolutionary Imagination

Droll thing life is—that mysterious arrangement of merciless logic for a futile purpose. The most you can hope from it is some knowledge of yourself—that comes too late—a crop of unextinguishable regrets.

—Joseph Conrad, *Heart of Darkness*

It is better for mankind to be impressionable than reflective. Nothing humanely great—great, I mean, as affecting a whole mass of lives—has come of reflection.

—Joseph Conrad, *A Personal Record*

Joseph Conrad is by far the most critically acclaimed turn-of-the-century novelist to engage with naturalistic cosmology. He was a Polish emigre who spent his childhood in Russian exile, his youth apprenticed to French sailors, and his middle age adulated by British intellectuals whose language he learned at 21. Were those factors overlooked, he would still be unique among the novelists discussed in this study. Unlike Arthur Conan Doyle, Robert Louis Stevenson, Edward Bulwer-Lytton, H. G. Wells, and Jack London, Conrad was not affiliated with any conventional ideological or moral system. Nor did he affirm the "inverted creed" of moral nihilism (Conrad 1996 [1912], 93). He was an agnostic with divided national allegiances (Najder 2011). In his own words, he was a resigned but sympathetic spectator of humanity: "open-eyed, conscious,

E. Jonsson, *The Early Evolutionary Imagination*, Cognitive Studies in Literature and Performance, https://doi.org/10.1007/978-3-030-82738-0_7

and informed by love" (Conrad 1996 [1912], xix). This elusive position was further veiled by his reticence about personal matters. As Virginia Woolf put it, "there clings to the genius of Conrad something essentially, and not accidentally, difficult of approach" (1984 [1925], 223). His taciturn cosmopolitanism makes him both a uniquely interesting case study of Darwinian influence and a difficult subject for any influence study. He was an exploratory intellectual, pointedly uninsulated by dogma, who approached the new human condition with the highest degree of psychological and aesthetic subtlety.

Conrad's familiarity with evolutionary theory has been as impossible to corroborate in detail as it has been to dismiss in general. A few of his novels unquestionably address naturalistic cosmology: *Almayer's Folly* (1895) with its sun-reaching plants "climbing madly and brutally over each other," *Heart of Darkness* (1898) with its "icthyosaurus"-haunted prehistoric backdrop, and *Nostromo* (1904) with its doomed "imaginative materialist" Decoud (Conrad 1992 [1895], 71; 1992 [1899], 45; Watt 1980, 44–45). Conrad's letters to the socialist MP and gaucho laird Robert Bontine Cunninghame Graham famously evoke the horror of the materialistic cosmos: "It evolved itself (I am severely scientific) out of a chaos of scraps of iron" (Conrad 2015, 81–82). These are not surprising references for an educated man of the day. In Ian Watt's words, "Conrad grew up in the heyday of evolutionary theory" (1980, 155). Despite his exile, he was a cultivated member of the Polish aristocracy, who claimed to be capable, already during his days at sea, of conversing "upon science, politics, natural history, and operatic singers" (Conrad 1996 [1906], 123; Najder 2011, 3–69). The difficulty lies in identifying specific sources for his knowledge.

Previous studies of Conrad's evolutionary thinking indicate that it is best to acknowledge the inconclusive evidence for his scientific knowledge. That inconclusiveness readily transforms into a license for speculation. At one extreme is Allan Hunter's *Joseph Conrad and the Ethics of Darwinism*, which argues that Conrad used his novels as scientific theses for "putting the record right about a large number of what he identifies as popular fallacies" (1985, 15). At the other extreme is George Levine's *Darwin and the Novelists*, which groups Conrad with authors like Dickens "who probably did not know any science first hand, who could have been 'influenced' by Darwin only indirectly" (1991, 3). These positions are equally speculative. Hunter has rightly been criticized for basing his claims for influence on parallel imagery and verbal associations

(Williamson 1985, 406; Hampson 1985, 262). Levine, in contrast, treats the absence of records as a record of absence. His claim that Conrad was insulated from scientific texts is hard to reconcile both with Conrad's identity as a late Victorian intellectual and with the information that does exist about Conrad's scientific interests.

In the face of Conrad's personal reticence, scholars have looked for illumination in his cultural environment. Ian Watt's *Conrad in the Nineteenth Century* (first published in 1979) best exemplifies that approach to his scientific knowledge. Watt describes a background of late Victorian scientific preoccupations and makes only modest claims for the influence of specific texts (1980, 45, 155–61). This cautious method has a different, inevitable downside: it portrays Conrad's scientific information as an average of his age's most visible ideas. No matter how complex Conrad's response to this average is taken to be, his concerns are defined by the mainstream concerns of late-Victorian culture, leaving out his significant idiosyncrasy. The same problem applies to the many explanations of Conrad's world view based on large-scale cultural movements. His friend and first biographer Richard Curle called him a "Realistic Romanticist" (Peters 2006, 119). Since then, scholars have appealed to different cultural terms along those two lines—from "disdainful skepticism" to "magazine-writer" romance, from "nihilism" to a Wordsworthian heritage that is "modern and Romantic simultaneously" (Mencken 1917, 16; Leavis 1950, 180; Peters 2006, 126; Thorburn 1974, 165). The late twentieth century has seen a shift of explanatory terms from aesthetics to politics and philosophy. A hundred years after Conrad's debut, his literary output is seen as a critique of "absentee capitalism," as a "laboratory of gender difference," as "hopelessly Anglophile and racist," or as a display of "the passionate irony of an exile" (Martin 2015, 584–86; Mallios 2013, 132; Moore 1996, 223, 231). Nevertheless, from aesthetic philosophy to political identity, the explanatory terms have been predominantly cultural.

Conrad's mind is less than usually reducible to an interaction between cultural influences. His pointed anti-dogmatism applies not only to religion, politics, and philosophy, but to aesthetic movements like Romanticism. His literary ideals range from Henry James's most high-strung psychological impressionism to the simplest naturalism of Ivan Turgenev and Guy de Maupassant, from Polish romantic poetry via all the modes of Gustave Flaubert to Shakespeare and the scientific romances of H. G. Wells (Conrad 2004, Najder 2011, 27, 47, 455–56). Though patterns may be seen among these influences—as among his influences from philosophy and politics—the influences themselves are not very illuminating as explanatory terms. It is clearly the case that Conrad combined resigned analytic detachment with imaginative absorption in the world. Even during his life, he was suspected alternately of an "unemotional, grim acceptance of facts" and of being "essentially sentimental and melodramatic" (Conrad 1996 [1912], xv; Wells 1934, 526). These two tendencies, on which scholars have converged using different cultural terms, can be significantly illuminated by the two levels of explanation that have tended to be left out: personality and universal human psychology.

By all accounts, Conrad was neurotic with depressive tendencies, sympathetic and highly imaginative, but also rigidly conscientious, emotionally restrained to an extreme, and intellectually humble (Conrad 2015; Galsworthy 1925; Najder 2011). That quite specific personality profile was modified by his cultural environment and life experience, but, as in the case of Jack London, it also shaped what he took in and how he expressed himself. Most notably, it shaped how he responded to one of the greatest psychological challenges of human history—a world robbed of mythology. I will piece together the information that does exist about Conrad's personal response to naturalistic cosmology, and analyze his world view on the three levels of individual identity, cultural context, and universal psychology. Conrad's famous skepticism can be explained as a way to balance his personality, stabilize his cultural identity, and accommodate the cosmic revelations of science. When those three levels come together, they form a life strategy to fulfill the human need for meaning.

Whether or not Conrad read Darwin, he displays the symptoms of assimilating Darwinian evolution imaginatively. His peculiar perspective contains patterns apparent in authors who were unquestionably preoccupied with such assimilation. Like Jack London and H. G. Wells, Conrad envisioned an essentially amoral naturalistic world in conflict with moral ideals. Like Stevenson, he saw the human mind divided against itself between contradictory impulses. Like Bulwer-Lytton, he engaged agonizingly with the relationship between human nature and political idealism. Like Doyle—albeit with differences in style, meaning, and magnitude—he depicted Western civilization encountering its deep past in tropical locales. The shared themes of these authors are partly explicable as universal human concerns. Like all humans, they needed to negotiate personal and social values in contact with traditions and the changing natural world, using minds prone to metaphysical dualism and simple agonistic structures. However, the particular shape of the themes is explicable as a response to post-Darwinian naturalistic cosmology. All of these authors responded to the unique challenges of a natural world alien to human justice, a moral world without transcendental guideposts, and a species-wide self-image that had no immutable essence.

When non-mythological cosmology is conceived as a unique challenge to the human imagination, and when the patterns of imaginative response are made clear, the individual differences among authors become more apparent. Conrad's skeptical distance allowed him to face transient humanity in an indifferent cosmos—to acknowledge the amoral natural forces without either infusing morality with spiritual qualities or rejecting the need for moral action. Neither nature nor culture appeared in this light as harmonious protagonist or vicious antagonist. But the distance also dwarfed his own actions and hopes. He clung to the tragic perspective in order to avoid satirical dismissal, and to humor and restraint in order to avoid despair and dissolution. In public, he declared that human affairs deserve "admiration and pity," "love" and "respect" (1996 [1912], xix). In private, he could admit "that to look at the remorseless process is sometimes amusing"—and he could burst out that it was "not worth troubling about," because "if you take it to heart it becomes an unendurable tragedy" (2015, 82, 89). Above "the dust of the conflict," he envisioned consciousness "powerless but concerned sitting enthroned like a melancholy parody of eternal wisdom" (quoted in Najder 2011, 315). The same pressures that turned Jack London into an oscillating superman socialist turned Conrad into a captive consciousness whose chief power was changing the genre of what it perceived.

Conrad's temperamental skepticism merged with his conflicting cultural alliances and the moral relativism revealed by science. His pragmatic imaginative sympathy and moral heritage made him see the reality of human social bonds and moral emotions as well as their fragility. With good reason, he has been connected to every major movement of post-Darwinian skepticism, from Existentialism to Postmodernism (Peters 2006, 128–33; Bohlmann 1991; Cox 1974; Erdinast-Vulcan 1999). With equally good reason, he has been credited with an aesthetically intense moral vision that harks back to the Romantics and aligns him with certain types of adventure fiction (Leavis 1950; Thorburn 1974; Woolf 1984 [1925]). His idiosyncratic combination of amoral cosmology and simple naturalistic morality represents a remarkable balance between human imaginative needs and non-mythological reality. That balance is powerfully enacted in his most famous work, which is also his most overtly evolutionary novel. An interpretive reading of *Heart of Darkness* can thus serve as a suitable capstone for this study of the late nineteenth-century evolutionary imagination.

7.1 Conrad's Science

It is worth emphasizing the high degree of uncertainty about Joseph Conrad's scientific education. Did he have a precise understanding of natural selection and an accurate concept of geological time? Such questions are doubly beset by scarce records and unreliable sources. Very little documentation survives from Conrad's early Polish years—or even, because of his mobility and the subsequent wars, from the first thirty-seven years of his life (Najder 2011, viii; Hampson 2012, 3). Much of his education was either supervised by his father or, later, pursued eclectically and through the recommendations of friends (Najder 2011, 27, 46–47; Conrad 2015, 84–244). His autobiographical writing is notoriously taciturn and unreliable—so obviously so that his first preface to *A Personal Record* busily defends his "restraint," and makes a dubious distinction between his *truth* and his *sincerity* (emphasizing "praiseworthy sincerity") (1996 [1912], xiii–xvi). He left no extended account of his reading. Despite this archival haze, it is clear that Conrad felt strongly about naturalistic cosmology. In the exhaustive biography of Zdzislaw Najder, embracing research across multilingual letters, Polish estates, and international shipping records, there are no index entries for *Charles Darwin, Darwinism, evolution*, or even *science* (Najder 2011). Yet, Najder affirms the scholarly tradition of connecting Conrad to Darwinism, and places the "scientific picture of the world" at the core of Conrad's world view (287, 253; Hunter 1985; Renner 1975; O'Hanlon 1984; Watt 1980). That interpretation is equally warranted in its certainty and its imprecision.

From the records that do exist, it is possible to reconstruct a fair amount of Conrad's personal attitude to naturalistic cosmology. His skeptical epistemology appears to have been connected to natural science. He was fascinated by the otherworldly imaginative impressions of physics, often using it to dwarf human strivings and frame life as tragic or absurd (Conrad 2015, 100–103; O'Hanlon 1984; Hampson 1985). According to his friend the novelist John Galsworthy, he was attracted to William James, with his pragmatic interpretations of coherence and belief, and to Arthur Schopenhauer, with his anti-teleological pessimism (1925). On this bleak cosmic screen, Conrad projected an idea of nature that was shaped from an early age by books of travel and exploration (Conrad 1996 [1912], 13: Griffith 1995; Hunter1985, 1; Tsao 2008, 46–47). His son John reports that "prehistoric animals had a great fascination for him," and that he would explain the reconstructed images in Henry Neville Hutchinson's *Creatures of Other Days* (John Conrad 2008, 18). One of his favorite travel writers was the co-discoverer of natural selection, Alfred Russel Wallace (Watt 1980, 155; Tsao 2008, 45). Another was the naturalist and novelist W. H. Hudson (Conrad 2015, 452; Miller 1990, 2; Najder 2011, 339). These nature chroniclers shaped Conrad's attitude to human life on an existential as well as an ethical level. He made Wallace a model for *Lord Jim*'s butterfly collector Stein, in famous dialogue with Marlow on how to live (Tsao 2008, 45). He praised Hudson for simultaneously seeing the beauty and the hunting instinct of a spider—and compared that perspective to the resigned social critique of a Galsworthy novel (Conrad 2010, 101–5).

One of Conrad's first British literary friends was H. G. Wells. Najder claims that Wells and Conrad found their momentary common ground through science: they were supposedly both "deeply interested in the foundations of human knowledge, the future of science, the theory of evolution, and cosmogony" (2011, 280). Indisputably, their most intimate period occurred when both writers treated science most actively in their novels (McCarthy 1986, 38; Dryden 2013, 214). Wells was publishing evolutionary allegories and visions of humanity's future: *The Time Machine* (1895), *The Island of Dr Moreau* (1896), and *War of the Worlds* (1897). Conrad was publishing *Heart of Darkness* (1898), *Lord Jim* (1899), *Nostromo* (1904), and an abortive scientific romance, co-authored with Ford Madox Ford, called *The Inheritors* (1901). Their friendship does not seem to have declined because they had different interpretations of naturalistic cosmology. Toward the end of their most intimate period, Conrad wrote Wells about the evolutionary political tract *Mankind in the Making* (1903). Conrad praises Wells's book but refuses to comment in detail: "Our differences are fundamental but the divergence is not great" (*Letters*, 165). He proceeds to draw two mysteriously tangling lines of "convictions," Conrad's and Wells's, and—when pressed—only objects to Wells's dogmatic rhetoric (*Letters*, 165–67). The problem seems to have been Wells's political attitude. During their last meeting, according to Hugh Walpole, Conrad revealed the mystery of those tangling convictions: "The difference between us, Wells, is fundamental. You don't care for humanity but think they are to be improved. I love humanity but know they are not" (quoted in Dryden 2013, 223).

Wells may have shaped Conrad's general understanding of biology, but he almost certainly intensified the image of the dead earth that haunted Conrad's mind. In 1896, Conrad notes that he has read *The Time Machine* and intends to remain a reader: "'Dr Moreau' is just now on his way to my island" (2015, 50). In 1900, the *Pearson's Magazine* article "How Will the World End?" by Herbert Fyfe outlines various scientific hypotheses about humanity's demise—hypotheses as different as celestial catastrophes, suffocation from industrial pollution (Lord Kelvin), and the atmosphere set ablaze by electricity (Nikola Tesla). The idea that life will freeze as the sun expires is attributed to George Darwin, but it is described (with rich illustrations) entirely through H. G. Wells's "strangely impressive picture" of it in *The Time Machine* (Fyfe 1900, 5). Two years after reading *The Time Machine*, Conrad evokes this specifically cold end of the world to Graham: "A moment, a twinkling of an eye and nothing remains—but a clot of mud, of cold mud, of dead mud cast into black space, rolling around an extinguished sun" (translated in 2015, 100). The cold Earth remained vivid enough in Conrad's mind to appear years later, somewhat incongruously, in his essay on Henry James: "When the last aqueduct has crumbled to pieces, the last air-ship fallen to the ground, the last blade of grass has died upon a dying earth, man... shall set this undiminished light of his eyes against the feeble glow of the sun" (2004 [1904], 16–17).

Conrad's cosmological reflections rarely include direct references to Darwinism. His imagery tends to proceed from astrophysics to ethics without the intervening step of biology. It is nevertheless clear that he considers human moral impulses subject to natural law. The mechanistic cosmos he describes to Graham is a horror version of Martin Eden's ecstatically unified universe, proceeding from physics to the human imagination: "It knits us in and it knits us out. It has knitted time space, pain, death, corruption, despair and all the illusions" (2015, 82). Another letter to Graham familiarly refers to humanity as "the victims of nature" and "part of the animal kingdom" (2015, 93). His autobiography paints conscience, with high rhetoric, as the joint product of biology and culture:

> Conscience, that heirloom of the ages, of the race, of the group, of the family, colourable and plastic, fashioned by the words, the looks, the acts, and even by the silences and abstentions surrounding one's childhood; tinged in a complete scheme of delicate shades and crude colours by the inherited traditions, beliefs, or prejudices—unaccountable, despotic, persuasive, and often, in its texture, romantic. (1996 [1912], 94)

Conrad's celebration of the "despotic" conscience is not out of keeping with his idea that humans are victims of a natural world both wonderful and cruel. It may not be an accident that his "heirloom of the ages" echoes the same passage from Tennyson's "Locksley Hall" that Wells subverts in *The Time Machine* when he reveals the monstrous Morlocks as descendants of humanity (Wells 2005 [1895], 46).

Perhaps Conrad's most overt reference to natural selection occurs in his essay on Henry James. He uses an extended metaphor of human life as a battlefield, which he defends with a connection to natural history:

> These warlike images come by themselves under the pen; since from the duality of man's nature and the competition of individuals, the life-history of the earth must in the last instance be a history of really very relentless warfare. (2004 [1904], 18)

This offhand remark locates humanity's history within the competitive history of all life. The "duality" Conrad sees in human nature is almost certainly egoism and altruism—two concepts that recur obsessively in his moral reflections (Conrad 2015, 92–93, 167; Najder 2011, 315). Unlike Stevenson, Conrad did not equate that duality with good and evil or connect it to a spiritual purpose. In a letter to *The New York Times* four years before his essay on James, he dwells on egoism and altruism as mankind's two "contradictory instincts" (quoted in Najder 2011, 315). Altruism is the "morality" of the world, but egoism is its "moving force," and neither can be a guide to human life; the only thing certain about them is "the incomprehensible alliance of their irreconcilable antagonism" (ibid.).

These fragments, taken together, suggest that Conrad incorporated scientific knowledge into his life strategy. Physics and biology mingled with literature and skeptical philosophy throughout a life of physical, political and financial turmoil, until he saw the cosmos as a great and terrible mechanism. Yet, he believed that even a spider deserved a double vision that acknowledged its beauty. Like Darwin—and unlike most of his contemporaries—Conrad successfully suspended his mind between the horror and wonder of nature. But Conrad did not have Darwin's temperamental equanimity, or his Victorian glimpses of moral instincts growing toward a future Utopia. Conrad's future was the cold Earth. His suspended mind was that of the concerned consciousness, committed to simple moral ideals that he expected himself and others to achieve only approximately and ephemerally: "sincerity, courage"; "honour, fidelity to truth in themselves and others" (Conrad 2015, 81, 90). This is an essentially tragic vision: as Najder puts it, Conrad "placed himself in the position of a soldier defending a doomed outpost" (2011, 256). His intellectual humility distanced him equally from the idea that mankind could be perfected and the idea that mankind was deplorable or unknowable: "what one feels so hopelessly barren in declared pessimism is just its arrogance" (2004 [1905]-b, 13). Naturalistic cosmology revealed a struggle of contradictory impulses, and Conrad's response was cosmic resignation coupled with unflinching human sympathy: "the courageous recognition of all the irreconcilable antagonisms that make our life so enigmatic, so burdensome, so fascinating, so dangerous—so full of hope" (2004 [1905]-b, 12).

7.2 The Skeptical Life Strategy

Conrad's skeptical resignation was a life strategy for the human world of thought: a way to organize his memories, actions, and aspirations, and to conceive of himself as a coherent self. It was a strategy at once deeply emotional and highly intellectualized. He was torn at every level of his identity: between the imaginative emotionality and restrained dutifulness of his personality; between the opposing political, national, and vocational traditions of his personal history; and between the irreconcilable amoral forces of naturalistic cosmology. He achieved consistency through his skepticism . Skepticism was not just a way to reconcile opposing commitments, but a basis for his own idiosyncratic moral attitude. He celebrated it in telling terms when he urged it on Galsworthy: "Scepticism the tonic of minds, the tonic of life, the agent of truth—the way of art and salvation" (2015, 146). From the intellectual distance of his skepticism, Conrad could see the positive experiences of life without trying to arrange them into unsustainable systems. He could satisfy his sympathetic nature by making sympathy global, and satisfy his need for conscientious restraint by moderating his emotional investment in the opposing forces of the world. Maintaining his skeptical position became a moral end in itself. His dedication to skepticism merged with his artistic ambition, with the simple social and vocational constancy he admired, and with the integrity of his own consciousness.

Scholarly interpretations of Conrad's moral attitude are tied up in the debates about his Romantic or Realist tendencies. Though perceptive critics sometimes seem to have seen different people when they looked at him, there are two distinct interpretations: Conrad as a moral relativist and Conrad as a simple moralist. Geoffrey Harpham suggests that Conrad was "simply incapable of achieving the commonplace," but F. R. Leavis calls him "in some respects a simple soul" (Harpham 1996, 102; Leavis 1950, 182). H. L. Mencken praises Conrad's "disdainful skepticism" and "wholesale negation of all morals," but Vladimir Nabokov groups his work with the morally simplistic "romantic productions" of Arthur Conan Doyle (Mencken 1917, 16; Moore 1996, 230). To Leavis, Conrad's moral relativism is mere "magazine-writer" rhetoric, which distracts from the "heroic sublimity" of his simple decent sailors (1950, 180, 185). To E. M. Forster, Conrad's worldview contained no philosophy and "no creed" (quoted in Leavis 1950, 173). These two broad interpretations of Conrad have persisted to the present. During the late twentieth century, J. Hillis Miller followed the line of moral relativism and sparked "an entire school of thought that saw Conrad approaching nihilism" (Peters 2006, 126; Miller 1965, 13–39; Cox 1974; Roussel 1971). Against this school, David Thorburn's *Conrad's Romanticism* championed a Conrad whose finest work carried the simple morality of Wordsworth into modernity (1974, 165). More recent scholarship has translated the same interpretations into the terms of contemporary literary study. Instead of Mencken's disdainful skepticism, there is proto-Existentialism, proto-Modernism, and proto-Postmodernism (Peters 2006, 128–33; Bohlmann 1991; Cox 1974; Erdinast-Vulcan 1999). Instead of Nabokov's romantic productions or Leavis's sublime heroism, there is a "Romance" set up to be reified, deconstructed, and turned toward political liberation (Peters 2006, 129; Erdinast-Vulcan 1999; Jameson 1982, 206–80).

Without understanding the unique challenge of naturalistic cosmology, it is difficult to account for Conrad's combination of moral relativism and simple morality. Both tendencies are essentially naturalistic: the moral relativism is a recognition that human morality is flexible and does not govern the cosmos; the simple morality is a recognition that humans are capable of moral values and social attachments, romantic love, devotion to crafts or ideas, and self-forgetful resilience. Because naturalistic cosmology does not relieve the moral relativism with any mythological purpose, the simple morality is psychologically necessary. For critics like Mencken, who explain Conrad's moral relativism through mythological types of pessimism—"Nirvana, the desire for nothingness, the will to not-will"—the simple morality is easily dismissed as ironic mockery (1917, 14). For critics like Leavis, who find in Conrad's simple morality a cosmically sufficient "triumph of the spirit," the moral relativism easily appears as irrelevant rhetoric (1950, 186). Only a few critics have described Conrad's peculiar double vision without dismissing either part as illusory or inferior. Virginia Woolf's obituary essay contains a lyrical description of his cosmos: people and places illuminated against the background of an incomprehensible world, summed up in "words of epitaph which remind us, with all that beauty and brilliancy before our eyes, of the darkness of the background" (1925, 227). Richard Curle noted Conrad's use of "irony and fixed ideas" (Peters 2006, 119). Galsworthy identified the source of those fixed ideas when he observed that Conrad mocked "all ideals and aspirations that are not grounded in the simplest springs of human nature" (1925).

Conrad's moral attitude would have been difficult to capture even without the complication of naturalistic cosmology, because it is an attitude—a way of thinking—rather than a system of principles. As Curle and Galsworthy noted, Conrad based his worldview on a "few very simple ideas": sincerity, fidelity, resignation or restraint, honor or conscience, and efficiency (Conrad 1996 [1912], xviii–xix). Those "simple ideas" are as nonspecific as Christian virtues, but without being organized by a mythological moral system. Translating them into action requires constant intellectual maintenance. Restraint to what extent? Efficiency to what end? There are no transcendental or even systematic answers. Thus, "Fidelity" was at different times embodied by literary art, seamanship, memory, civilization, the craft of yachting, love, Poland, and Lord Nelson as the British national spirit (2004 [1904]; 1996 [1912], xix; 1992 [1899]; 1996 [1906], 3–13, 23–34, 183–94). That type of moral attitude becomes simply invisible to critics like E. M. Forster, who look for a set of unbending principles:

> We needn't try to write him down philosophically, because there is, in this direction, nothing to write. No creed, in fact. Only opinions, and

> the right to throw them overboard when facts make them look absurd. Opinions held under the semblance of eternity, girt with the sea, crowned with stars, and therefore easily mistaken for a creed. (quoted in Leavis 1950, 173)

Conrad did modify his opinions in contact with facts, and he subjected every moral position to skeptical scrutiny. But the coherence of his moral attitude is not a false impression produced by aesthetics. From craftsmanship to nationalism, his disparate fidelities have a common aim: to preserve a coherent consciousness, and conscience, against an amoral and ephemeral world. The goal of Conrad's exploratory moral attitude is sincerity about the self and the world, fidelity to a way of thinking and feeling, honorable restraint toward others, and efficiency in interacting with the world. Though this moral attitude follows his temperament by emphasizing intellect and self-control, the ideals that govern it are rightly framed—by Conrad and Galsworthy—as universal human concerns.

Conrad's moral ideals are so general that their distance from contemporary social controversies seems almost geological. He aptly describes them as "so simple that they must be as old as the hills" (1996 [1912], xix). Though he knew that the hills were no more permanent than the sun, values that stretched across human history provided him with some psychological stability. Conrad's ultimate fidelity is to the skeptical distance that maintains his stability. Considering the threats to his personal, cultural, and cosmological coherence, the importance of that fidelity can hardly be overstated. One letter to Graham makes the point with particular force. Conrad refuses to give a speech at Graham's political meeting, because he needs to keep his "thinking inviolate" as a "final act of fidelity to a lost cause":

> It's all I can do. I have thrown my life to all the winds of heaven, but I have kept my way of thinking. It's a little thing—it's everything—it's nothing—it's life itself. This letter is incoherent, like my life, but the highest logic is there nevertheless—the logic that leads to madness. (translated in Conrad 2015, 116–17)

This is not the statement of a moral relativist or a simple moralist, but of a skeptic who felt the transience of simple attachments and still rested his world on them. Conrad's resistance to Graham's political program did not mean alignment with any other political program. He registered the absurd range of contradictory moral positions and remained invested in a moral attitude that, for him, amounted to an inviolable creed.

How did Conrad form his peculiar cosmic vision, so different from those of Jack London and H. G. Wells, Arthur Conan Doyle and Robert Louis Stevenson—or even those of Arthur Schopenhauer and Charles Darwin? His own explanation to Graham characteristically mingles the personal and the culturally specific with the cosmological. Though the letter begins by discussing a particular political meeting, Conrad's reflections immediately become so general—passing through the moral ambivalence of society, universal fraternity, nationhood, and selfishness—that there are no longer any single referents. The "lost cause" is the independence of Poland and the revolutionary idealism of his martyred parents, but also the coherence of his personality and his moral attitude as a whole; the "way of thinking" is his attachment to Poland and the anti-revolutionary attitude of the uncle who raised him, but also the skeptical restraint that sustained his identity; the "highest logic" is human nature and the mechanistic transience of the cosmos, but also the state of European politics and the course of his own life (Conrad 2015, 113–17). Though none of these explanatory levels can stand alone to account for Conrad's cosmic vision, disentangling them illuminates its peculiar character.

The Personal Level

Many who met Conrad noted his contradictory personality traits: his combination of restraint and imaginative volatility. Virginia Woolf's obituary essay summarizes the reports of visitors in two telling superlatives: "the most perfect manners" and "the brightest eyes" (1984 [1925], 223). Henry Newbolt, seeing Conrad for the first time, described him as "commanding" and exuding "intellectual calm" but gradually revealing "an artistic self, sensitive and restless to the last degree" (quoted in Najder 2011, 331). During a conversation, as Newbolt reports, Conrad startled his hearers by imitating the effect of the London crowd on his nerves: "I see their personalities all leaping out at me like *tigers!*" (ibid.). The next moment, the tiger talked "wisely and soberly as if he were an average Englishman with not an irritable nerve in his body" (ibid.). Galsworthy, from a more intimate perspective, claimed that Conrad suppressed "an almost ferocious enjoyment of the absurd": his sense of humor is supposedly powerfully restrained in his novels and "dry and sardonized" in his letters, but would in conversation "leap up in the midst of gloom or worry, and take charge with a shout" (1925). H. G. Wells spoke of Conrad's "over-sensitized receptivity," but also of his "nautical trimness" and his "singleness and intensity of purpose that made the kindred concentration of Henry James seem lax and large and pale" (1934, 525, 526, 528). Conrad's son John paints a domestic portrait of his father that reveals patterns of anxiety, sympathy, excitability, imaginative play and ritualized self-control (2008). According to John Conrad, the habit of self-control was so pronounced that his father hurriedly sent him out of the room after autographing *The Rescue* on request: "At no time did he like to show emotion but on this occasion I believe he came very close to it" (2008, 168).

Conrad's need for self-possession may at times seem comically exaggerated, but he was managing a highly sensitive personality put under pressure through trauma. He had experienced the loss of both parents and several homes before puberty. The most extreme threat to his self-possession was the paralysis of his mind through depression. His periods of depression appear to have followed him from early childhood to the last decades of his life (Najder 2011, 46, 66, 165, 167–68, 418–19, 577). During his youth in Marseille, in the midst of murky social and financial affairs, he attempted suicide (Najder 2011, 66–67). His letters to close friends are full of testimonies to these depressive periods, such as the chilling aside to Edward Garnett in 1899:

> Fact is I am not worthy to take up your thought. The more I write the less substance do I see in my work. The scales are falling off my eyes. It is tolerably awful. And I face it, I face it but the fright is growing on me. My fortitude is shaken by the view of the monster. It does not move; its eyes are baleful; it is as still as death itself—and it will devour me. Its stare has eaten into my soul already deep, deep. I am alone with it in a chasm with perpendicular sides of black basalt. Never were sides so perpendicular and smooth, and high. Above, your anxious head against a bit of sky peers down—in vain—in vain. There's not rope long enough for that rescue. (Conrad 2015, 122)

Conrad's restraint was a way to manage rather than repress his volatility. As Najder suggests, his verbal art may have been partly "psychotherapeutic" (2011, xii). Conrad was constantly giving imaginative shape to the strong impressions made on him by the world—the tigerish personalities of the Londoners or the baleful monster of his literary failures. These images, like the images in literature and mythology, turned inchoate emotion into distinct shapes that could be composedly approached. When his son came crying after seeing a face in a tree, Conrad did not tell him to suppress the fantasy, but took him by the hand and made him "show him the face, pointing out the eyes, nose and mouth with a stick" (John Conrad 2008, 20). He peopled his son's mind with living trees because, as he saw it, the boy needed to "get used to having an imagination and not be frightened by it" (20). Conrad's own fears were not as easily mastered, but he used the same approach. His autobiography casts him as "an historian of hearts" who reaches composedly for "the very fount of laughter and tears" (*Personal*, xxi, xix).

Like most people, Conrad generalizes to a certain extent from his own personality. In the preface to his autobiography, he acknowledges that his restraint is in one light "merely temperamental" (1996 [1912], xvi). He admits to a "perfect horror" of losing his self-possession, and attributes it to "pride": "There can be nothing more humiliating than to see the shaft of one's emotion miss the mark of either laughter or tears" (1996 [1912], xvii, xvi). Despite this awareness, he elevates restraint into a universal human virtue and declares resignation "the only one of our feelings for which it is impossible to become a sham" (1996 [1912], xix). He even rationalized his own dislike of emotional display as a commitment to sincerity:

> As in political so in literary action a man wins friends for himself mostly by the passion of his prejudices and by the consistent narrowness of his outlook. But I have never been able to love what was not lovable or hate what was not hateful out of deference for some general principle. … I have always suspected in the effort to bring into play the extremities of emotions the debasing touch of insincerity. (1996 [1912], xvii)

The fact that Conrad's personality predisposed him to skepticism saved him from a certain narrow certainty in universalizing that predisposition. He was quick to reinstate intellectual humility, to note that he did not "think resignation the last word of wisdom" and that his moral ideals were only his personal "conviction" (1996 [1912], xix). In a moment of characteristic self-irony, he captures his predisposition to skepticism; he is too disgusted by emotional partisanship to be either intensely emotional or fully composed: "imperfect Esthete, I am no better Philosopher" (1996 [1912], xx).

The Cultural Level

When reconstructing his personal history, Conrad faced a set of claims that would have taxed a person less concerned with integrity: a homeland abandoned in its extremity for two separate international vocations, two adoptive nations in which he had—respectively—come of age and married, and two father figures who were ideologically and temperamentally opposed (Najder 2011). He was born to revolutionary patriots and exiled along with them at five. After losing both parents, he was raised by an uncle who opposed Polish revolution. He left Poland for Marseille to go to sea against his uncle's wishes, and left ships and France for an English literary career. This was the personal history that made him tell Graham he had thrown his life "to all the winds of heaven" (2015, 116). His personal need for stability and coherence meant that he retained strong attachments to all the opposing claims of his personal history. His skeptical distance helped him suspend those opposing claims without dissolving into incoherence. The simple ideals of his moral attitude gave him direction, but were so general that they could contain his two father figures, his three home countries, and his two vocations.

It would have been miraculous if Conrad's family history had not left deep tracks in his mind. His mother Ewa Bobrowska was killed by the Russian climate when he was eight, and his father Apollo Korzeniowski followed three years later. Apollo was a figure of national fame who shared his son's imaginative neuroticism: a temperamental poet, translator, playwright, and radical political activist, immersed in Catholicism and Romantic nationalism. Around the time of Apollo's death, the eleven-year-old Conrad reportedly wore a revolutionist cap and wrote plays for his schoolfellows about "insurgents fighting against the Muscovites" (Najder 2011, 33). At twelve, Conrad became the ward of his uncle Tadeusz Bobrowski, who lived by ideals of restraint, industriousness, clear thinking, and financial responsibility (Najder 2011, 36–37). Tadeusz canonized Conrad's mother and treated Conrad with affection, but also chided him for dreamy irresponsibility and frequently implied that his "shortcomings were characteristic of his father" (Najder 2011, 36). Conrad came to love his uncle and be grateful for his influence; he admired Tadeusz's "powerful intelligence and great force of character" (Conrad 2015, 138). But he also considered his parents symbols of noble sacrifice, and remained emotionally invested in the cause of Polish independence (Najder 2011, 4–35, 529, 571–72). When he describes the thousands of mourners lining Apollo's funeral procession in Cracow, he calls it "the most natural tribute in the world—not to the man but to the Idea" (1996 [1912], viii; Najder 2011, 35).

Through his family history, Conrad had experienced the transition from revolutionary rhetoric to lethal consequences and personal grief. His revolutionist parents had died and his appeaser uncle had lived. This no doubt strengthened his view of political crusades as futile turmoil. In the introduction to his autobiography, he takes a stance against political agitation that approaches intellectual parricide. He calls rhetoric a stimulus for emotional excess and insincerity: "a fool's bargain," like selling one's soul "for love or power to some grotesque devil" (1996 [1912], xvii). His stance against "the revolutionary spirit" is among the least ambiguous he took: "Its hard, absolute optimism is repulsive to my mind by the menace of fanaticism and intolerance it contains" (1996 [1912], xx). This stance was not just public. Conrad upheld it against his intimate correspondents whenever they showed signs of revolutionary rhetoric. He chided Wells for writing "only for people who think this or that," Galsworthy for hugging his "conceptions of right or wrong too closely," Bertrand Russell for believing in "the advent of international socialism," and Graham for setting out to change "human nature" rather than institutions: "Your faith will never move that mountain" (Conrad 2015, 90, 146, 167, 456). It is, however, significant that so many of his intimate correspondents required chiding.

Conrad's polemics against his idealist friends were reliably accompanied by expressions of sympathy for their moral positions. He rejected Russell's vision for a socialist council in China because he thought it required an unrealistic "change of hearts": "There is not enough honour, virtue and selflessness *in the world* to make any such council other than the greatest danger" (2015, 457–58). He told Wells to take "sufficient account of human imbecility," and to avoid "an exclusiveness of feeling which (legitimate as it may be) can only serve your sincerity at the expense of truth" (2015, 166). Reliably, he was attracted to the moral aspirations of his friends and repulsed by their claims that human morality could be perfected through a particular scheme. He brought the issue to a head with Graham, whose aristocratic family, egalitarian politics, and nationalism gave him a more than passing resemblance to Apollo:

> You are a most hopeless idealist—your aspirations are irrealisable. You want from men faith, honour, fidelity to truth in themselves and others. You want them to have all this, to show it every day, to make out of these words their rule of life. The respectable classes which suspect you of such pernicious longings lock you up and would just as soon have you shot—because your personality counts and you can not deny that you are a dangerous man. *What makes you dangerous is your unwarrantable belief that your desire may be realized.* This is the only point of difference between us. I do not believe. And if I desire the very same things no one cares. Consequently I am not likely to be locked up or shot. Therein is another difference—this time to your manifest advantage. (2015, 82, my italics)

In the idealistic aspirations of his friends, all of which he considered unwarrantable, Conrad saw threats of the turmoil with which he was personally familiar. His "très cher ami" Graham was dangerous in a double sense: to himself (he had actually been jailed during a demonstration), and to the stability of British society.

Social stability was a guiding concern for Conrad, as a prerequisite for calm consistency and mental freedom. This concern had more to do with stability than with any particular society. His triple national identity aligned him with European civilization in broad terms: he defined Poland as "Western in complexion," with "its tradition of self-government, its chivalrous view of moral restraints and exaggerated respect for individual rights" (1996 [1912], vii). Both this national identity and his skeptical life strategy made him recoil from autocracy; but he did not want revolution for democracy in Russia and China any more than he wanted revolution against it elsewhere (Conrad 2004 [1905]-a; 1911; 2015, 455–59). His dislike of political turmoil produced statements resonating with opposite ends of the political spectrum. One letter to Graham condemns the socialist movement in Poland "because its propaganda (the only thing really tangible about it) tends to weaken the national sentiment the preservation of which is my concern" (2015, 114). Another letter to Graham condemns the jingoism surrounding the Anglo-Boer war: "All that's art, thought, idea will have to step back and hide its head before the intolerable war inanities... If I am to believe Kipling this is a war undertaken for the cause of democracy. C'est à crever de rire" (2015, 126). The common element in these statements is a distaste for destabilizing rhetoric.

Conrad's moral attitude was genuinely cosmopolitan and apolitical, though it distinguished between more and less admirable forms of behavior. His distaste for the revolutionary spirit did not prevent him from portraying it with understanding and partial sympathy in *Nostromo* and *Under Western Eyes*—both of which ultimately satirize it piercingly. His childhood experience of Russia did not prevent him from being horrified by the suffering on both sides of the Russo-Japanese war, or from admiring the Russian police captain ("this good man") who behaved decently when his dying mother was returned to exile (2004 [1905]-a; 1996 [1912], 65–66). Nor did he hesitate to call Russian autocracy "the brutal destruction of dignity, of truth, of rectitude, of all that is faithful in human nature," or to repudiate the Russian Governor-General who gave the order to exile a dying woman (2004 [1905]-a, 82; 1996 [1912], 67).

As Harpham puts it, Conrad's cultural experience was "comprehensive," ranging from nationless orphanhood to intellectual mastery within the British Empire: "a fact that confuses the moralizing mind, and is, indeed, genuinely confusing" (1996, 10). But Conrad's moral attitude is not incomprehensible if one steps outside clear-cut cultural affiliations. His appreciation for Ivan Turgenev as an "impartial lover of *all* his countrymen" parallels his appreciation for Juliusz Slowacki as " *l'âme de toute la Pologne*" [the soul of all of Poland] (2004 [1917], 42; quoted in Najder 2011, 459). His imaginative sympathy impelled both his recoil from social upheaval and his penchant for those who wanted to improve the world. When, in 1910, he signed a memorial urging the Prime Minister to support the Women's Suffrage Bill, he privately commented, "with the greatest sympathy for the object I cannot share the beautiful optimism of the memorial" (2015, 268). This comment sums up the political position of a skeptic committed to simple moral ideals—an exploratory mind expecting to be thwarted but refusing to become rigidly pessimistic. As he told Graham, "Everything is possible, and most things come to pass (when you don't want them)" (2015, 80).

The Universal Level

At the highest level of generality, Conrad's moral attitude was a response to naturalistic cosmology. His "way of thinking"—the moral relativism, the restrained but acknowledged emotions, the global sympathy and refusal of moral systems, the commitment to simple human attachments—was also his way of approaching the cosmos. It was a highly conscious approach. He knew, as he put it to Galsworthy, that his way of thinking meant both "everything" and "nothing"—everything on the human level and nothing on the cosmic. The discrepancy between those two levels was not to Conrad a cause for resentment or revolt, as it was to Jack London, but a subject for tragicomic contemplation tending toward the tragic. His mind, like most human minds, was incapable of remaining unflinching in face of the aimless suffering of naturalistic life. He achieved a balance between empathetic immersion and detachment by seeing the cosmos as a vast tragicomedy, alternately emphasizing his opposing emotional responses to it: elevating grief at its transient passions and absurdist amusement at its futility.

Conrad united the human level and cosmic levels by seeing humanity as the appointed spectators of the cosmos. In his autobiography, he openly rejects the idea that the universe is guided by human justice. As he puts it, trying to discern an ethical purpose of the universe "involves us at last in so many cruel and absurd contradictions" that he has come to suspect that "the aim of creation cannot be ethical at all":

> I would fondly believe that its object is purely spectacular: a spectacle for awe, love, adoration, or hate, if you like, but in this view—and in this view alone—never for despair! Those visions, delicious and poignant, are a moral end in themselves. The rest is our affair—the laughter, the tears, the tenderness, the indignation, the high tranquility of a steeled heart, the detached curiosity of a subtle mind—that's our affair! (1996 [1912], 92)

This is perhaps the most overt statement of Conrad's peculiar double vision. The cosmic level—"purely spectacular"—has no purpose beyond its existence. The human level—"our affair"—includes the ideals on which he rested his world: every bond or vocation resulting in sorrows and joys, every moral, aesthetic, and exploratory pleasure of the mind. But the spectacle of the cosmos itself is also our affair: "the unwearied self-forgetful attention to every phase of the living universe reflected in our consciousness may be our appointed task on this earth" (1996 [1912], 92). Conrad escaped the despair of crumbling mythologies by inventing a mythology of pure observation, adaptable—at least in theory—to whatever was observed.

Conrad's idea of humanity as cosmic spectators is essentially equivalent to the skeptical resignation that guided his own life. He himself wanted to observe life with open-eyed detachment informed by love, and he wanted humanity to observe the "living universe" with steeled but wakeful hearts. The ideals for both types of observation are the same: fidelity to a way of thinking and feeling, sincerity about the self and the world, restrained global sympathy. Human spectators should be guided by "conscience," which inspires "true testimony" of the universe, and they should rise unflinchingly above private emotions: "detached, impersonal," "self-forgetful" (1996 [1912], 92). Yet, Conrad found particular comfort in his cosmic spectacle because it acknowledged the emotional needs of humanity. Supposedly, a universe with no purpose beyond its existence favored mythologies that eased human existence. Hence, Conrad made room in his non-ethical universe "for every religion except for the inverted creed of impiety, the mask and cloak of arid despair; for every joy and every sorrow, for every fair dream, for every charitable hope"; he claimed that his cosmic vision provided "rational existence" for "inspiration of every sort" (1996 [1912], 93). Beliefs were good to the extent that they lead to healthy lives in sincere and coherent interaction with the world. As he put it elsewhere, "all ambitions are lawful except those which climb upward on the miseries or credulities of mankind" (1996 [1912], xviii). Through his amoral cosmic purpose, he arrived at a form of tolerant mythological pragmatism. He tried to counter the non-human cosmos by declaring that nothing cosmic was alien to him, inside or outside the human mind.

Conrad's cosmic vision was based on his personal temperament, and clearly favored his personal temperament. Though he expressed respect for most types of faith, he invested ultimate meaning in the position of skeptical observation that recognized the amoral nature of the universe. In Galsworthy's words, "He stared life very much in the face, and distrusted those who don't" (1925). The personal ideals from which Conrad generalized were based on his particular combination of imaginative volatility and stoic resignation. They also centered on literary expression. Supposedly by stray association, he notes that his spectacular universe accords a special place to "the artist of every kind" (1996 [1912], 93). In fact, something might be said for that natural witness to the universe, the novelist:

> Even he, the prose artist of fiction, which after all is but truth often dragged out of a well and clothed in the painted robe of imaged phrases—even he has his place amongst kings, demagogues, priests, charlatans, dukes, giraffes, Cabinet Ministers, Fabians, bricklayers, apostles, ants, scientists, Kaffirs, soldiers, sailors, elephants, lawyers, dandies, microbes and constellations of a universe whose amazing spectacle is a moral end in itself. (1996 [1912], 93)

Conrad admits that there are many people for whom neither self-analysis nor cosmological consciousness is a main concern, but simply concludes that it is "a pity": they are "unlucky" to miss humanity's "true task" (1996 [1912], 92). Not by chance, his cosmic vision is identical to the philosophy he urged on writers of fiction: to observe "the things of this earth" with understanding and love, neither submitting to "declared pessimism" nor measuring them against "some heaven of perfection" (2004 [1905]-b, 14). This double vision is the same that he admired in Hudson's portrait of the spider and Galsworthy's portrait of society.

By avoiding ideas of cosmic good and evil, Conrad held his world together against the metaphysical dualism so common among his contemporaries. In his most generous view—publicly declared—good and evil are relatively minor variations in the behavior of humanity. Good is its "obscure virtues" and evil its "small failings," "errors," "prejudices," and "platitudes and inanities" (2004 [1905]-b, 13). In his least generous view—played up in polemics with Graham—the universe is a non-moral "tragic accident" with a vaguely malicious tint: "the most withering thought is that the infamous thing has made itself; made itself without thought, without conscience, without foresight, without eyes, without heart" (2015, 82). In both views, there is only one universe. Though Conrad spoke locally of "fate" and "creation"—and, according to his son, once made a religious visit to an English church—he did not express any sustained belief in a separate spiritual realm (1996 [1912], 92; John Conrad 2008, 152). Arguing against Graham's political idealism, he observes that the cosmic machine cannot be improved by "celestial oil": "You cannot by any special lubricant make embroidery with a knitting machine" (2015, 82). Even the human consciousness in which he invested ultimate meaning was to him "merely the most enduring of the things of this earth, imperishable only as against the short-lived work of our industrious hands" (1996 [1912], 92; 2004 [1904]).

The lack of spiritual relief differentiates Conrad's cosmic vision from both Jack London's and H. G. Wells's. Conrad's moral ideals were not associated with souls and grails, and they were not conceived as the path to a better world. His skeptical resignation, fed by temperament and experience, prevented him from declaring war on the cosmic order. He occasionally admits that he wishes to "smash" the cosmos, or expresses respect for the "extreme anarchists" who want general destruction, but even such pessimistic outbursts end on a note of acceptance: "It has happened. You can't interfere with it... It is what it is—and it is indestructible!" (2015, 82, 116). To relieve the gloom, he retreats to his spectator's position and admits that the cosmic spectacle ("the remorseless process") is "sometimes amusing" (2015, 82). But detachment alone could not banish his cosmic horror.

Skeptical resignation is a position vulnerable to the paralysis of thought and action. *Nostromo* contains a sobering depiction of that danger: Martin Decoud descends toward suicide through "the sadness of the sceptical mind," which breaks up his world until he sees it "as a succession of incomprehensible images" (1974 [1904], 409). His death is brought on by the very open-eyed observation that Conrad favored. Decoud is "a victim of the disillusioned weariness which is the retribution meted out to intellectual audacity" (412). Like Jack London, Conrad occasionally lashed out at thought itself for subjecting him to this danger:

> The attitude of cold unconcern is the only reasonable one. Of course reason is hateful—but why? Because it demonstrates (to those who have the courage) that we, living, are out of life—utterly out of it. The mysteries of a universe made of drops of fire and clods of mud do not concern us in the least. (2015, 89)

Conrad's simple moral ideals acted as a counterweight to this cosmic horror—in part through the sheer distraction of pursuing them. In a letter from Africa, before his literary career, he celebrates the therapeutic power of sincere efficiency: "the forgetfulness and peace of work that is creative and absorbing" (2015, 16). When, as a literary author, he imagines the ideal life for his son Borys, the opposition between thought and work is even more marked: "I would like to make a bargeman of him: strong, knowing his business and thinking of nothing. That is *the* life" (2015, 96). But distraction is by definition a temporary solution—and, indeed, it fails Decoud when he is alone outside society. The long-term cosmic comfort of Conrad's moral attitude was that it centered on cross-cultural human values without needing to have them supported by cosmic laws. If the purpose of the universe is to be a spectacle for humanity, there is reason to prefer the human behaviors that leave a lasting favorable impression in the minds of human spectators. Fidelity to people and tasks has value simply because many of us consider it valuable. In this convoluted way, Conrad's mythology invested meaning in human attachments on the human level.

Conrad's skeptical management of his emotions reached all the way to the cosmic level. He refused to become overly immersed either in the tragic or the comic perception of humanity's place in the cosmos. Decoud's suicide comes partly of too much ironic detachment—he plays "lightly a part in a tragic farce" and looks "upon its terrible humours with amusement" (1974 [1904], 302). Decoud's friend Charles Gould, a practical idealist, despairs for the opposite reason: "It was tragic enough for him in all conscience, but he could see no farcical element" (302). Conrad himself perpetually saw the tragedy and the farce, often bewilderingly intermingled:

> It is very difficult to be wholly joyous or wholly sad on this earth. The comic, when it is human, soon takes upon itself a face of pain; and some of our griefs (some only, not all, for it is the capacity for suffering which makes man august in the eyes of men) have their source in weaknesses which must be recognized with smiling compassion as the common inheritance of us all. Joy and sorrow in this world pass into each other mingling their forms and their murmurs in the twilight of life as mysterious as an overshadowed ocean, while the dazzling brightness of supreme hopes lies far off, fascinating and still, on the distant edge of the horizon. (1996 [1912], xvi)

The "hopes" at the edge of the Conradian world plausibly represent idealism in the most general sense. In his essay "Books," he declares that "to be hopeful in an artistic sense" means believing that "there is no impossibility" of the world being made good (2004 [1905]-b). In his essay on Henry James, he imagines that the last witness at the end of the earth will express "some hope now to us utterly inconceivable," because of the human capacity to expect victory even when defeated (2004 [1904]). This was the capacity he simultaneously admired and mocked in his idealist friends.

Though Conrad did not subscribe to any idealistic hopes of his own, he portrayed such hopes as a light beyond ordinary human emotions. He clearly expected the last cosmic spectator to perish disproved, but he saw that spectator's expression of hope—like other expressions of faith and inspiration—as valuable in itself. It was what he, as a watching consciousness, preferred to watch:

> Mankind is delightful in its pride, its assurance, and its indomitable tenacity. It will sleep on the battlefield among its own dead, in the manner of an army having won a barren victory. It will not know when it is beaten. And perhaps it is right in that quality. The victories are not, perhaps, so barren as it may appear from a purely strategical, utilitarian point of view. (2004 [1904])

Idealistic hopes had a destabilizing power, and they could lead to the dreaded dogmatism of the revolutionary spirit, but they were also the opposite of depressive paralysis—and of Decoud's fatal detachment. Instead of using socialism as a handcuff to life or creating a World State in his mind, Conrad held onto the idea that humans can hope.

Though Conrad was suspended between the tragic and the comic perspective, he undeniably tended toward the tragic. He saw the suffering behind every farce but could not bring himself to smile at every grief. Humor seems to have provided him with relief on both the personal and the cosmic level. After a conversation with a physicist, he told Garnett that he hoped he could get a new house for his family: "If I don't I shall vanish into space (there's no space) and the vibrations that make up me, shall go to the making of some other fool" (2015, 103). His comic perspective could be ruthless, as in his exchanges with Graham about the knitting machine. But it was in his tragic mode that he faced the full significance of human life. Perhaps the most concise statement of his tragic vision appears (with characteristic peculiarity) in an essay on yachting:

> Love, though in a sense it may be admitted to be stronger than death, is by no means so universal and so sure. In fact, love is rare—the love of men, of things, of ideas, the love of perfected skill. For love is the enemy of haste; it takes count of passing days, of men who pass away, of a fine art matured slowly in the course of years and doomed in a short time to pass away, too, and be no more. Love and regret go hand in hand in this world of changes swifter than the shifting of the clouds reflected in the mirror of the sea. (1996 [1906], 25)

This tragic vision depends on the simultaneous recognition that human attachments are valuable and ephemeral. Unlike the spiritual Romantics, Conrad concludes his tragedy with irreversible personal and global destruction. Unlike the pessimistic Realists, he celebrates simple values that have been upheld across cultures throughout time—the love of men, of things, of ideas, and of perfected skill. That particular double perspective, along with his minimal personal mythology, sets his cosmic vision apart among the late-Victorian engagements with evolution. His peculiar cosmic vision pervades the novel in which he deals most directly with the challenge of amoral nature: *Heart of Darkness*.

7.3 *Heart of Darkness* (1899)

Conrad's most famous novel, like his general outlook, merges the personal with the cosmic. *Heart of Darkness* is based on his own experience in the Belgian Congo, and it is an allegory of the human mind encountering amoral nature. The novel is structured as a dream-like retelling of a journey to Africa by the master mariner Charles Marlow. His whole story is told between ebb and tide as dusk falls on the deck of a cruising yawl on the river Thames. The events of the journey are few and relatively simple: Marlow takes employment as a steamboat captain for a Belgian company trading in Ivory in the Congo. He travels to Africa, finds his steamboat in disrepair, repairs it, and follows the river to retrieve the sick ivory agent Kurtz. Kurtz has become a depraved warlord and dies on the way back with an expression of horror on his lips. Marlow then falls ill himself, returns to Brussels, and tells Kurtz's fiancée that Kurtz died pronouncing her name. The imaginative effect of the novel depends on Marlow's reflections on his experience. He begins his tale with a parallel between the Congo river and the river Thames in early Roman Britain, then weaves an ever-tightening connection between his journey and the fate of human civilization. He observes the senselessness and cruelty of the Belgian ivory trade, hears rumors of Kurtz's brilliant idealism and becomes psychologically allied with him, then recoils violently from the real Kurtz. Despite his recoil, he affirms a form of moral victory in Kurtz's dying words, and in speaking to Kurtz's fiancée, he helps sustain her illusion about Kurtz's moral nobility. As the story progresses, Marlow increasingly struggles to convey his experiences and make sense of them, while the novel's first narrator—one of Marlow's listeners—puzzles over both the story and Marlow himself.

Heart of Darkness is difficult even for its narrators to interpret. For Conrad, that difficulty was central to the novel's meaning. Marlow is faced with the extremes of human moral depravity and sees his society—and all of civilization—as a flicker in the amoral natural world. Like Conrad, he presents that experience to his audience without spiritual relief. He does not filter it through moral systems or ideas of supernatural justice. Marlow quietly judges the behavior of himself and those he encounters, admiring restraint and constructive effort, despising self-serving cruelty, cowardice, and loss of control. But these simple ideals are not enough to insulate him from the cosmic senselessness. Kurtz's reputation as a great idealist gives Marlow hope of some higher interpretation. Through Kurtz's depravity, that hope becomes cruelly ironic. The capacity for faith has turned foul in the wilderness, producing a greater force of destruction than the ordinary greed of the other ivory traders. But when Marlow lies to Kurtz's fiancée, half in spite of himself, he ends up on the side of that capacity for faith. Though he does not himself have a system of beliefs, he prefers to watch those who do. Like Conrad, he is ambivalently attracted to the supreme hopes of humanity. And like Conrad, he himself places faith only in the coherence of his own consciousness, held together by simple moral ideals. The novel comes to rest in a position of cosmic observation that acknowledges the full range of contradictory human behaviors, including the impulse to interpret the cosmos and to believe that it is meaningful.

In a critical history of *Heart of Darkness*, Ross Murfin considers it beyond doubt that the novel is "about the immorality of whites in Africa" (100). This focus on social issues has been shared by scholars who interpret the novel's politics very differently. Most critics up until the mid-twentieth century thought that Conrad predominantly sympathized with the Congo natives and critiqued the Belgian colonists (Murfin 1992, 100; Watts 1996, 52). In the late 1970s, Nigerian novelist and professor Chinua Achebe famously denounced *Heart of Darkness* as racist on the grounds that it centers on the immoral whites (Achebe 1977; Moore 1996, 127). Achebe's critique had more to do with the outgroup perspective of the dominant West than with Conrad's novel—and, as Gene Moore suggests, it is both rhetorically understandable and ironic that he targeted an author largely resistant to nineteenth-century racial prejudice (1996, 225, 231–33). But Achebe's political case against *Heart of Darkness* inspired a vast critical tradition (Watts 1996, 52–53; Peters 2006, 127). Because Conrad cannot be tied to any consistent political position, he has been freely appropriated on both sides of contemporary debates about class, gender, and ethnicity (Moore 1996; Murfin 1992). Hence, Peter Mallios and Gabrielle McIntire both consider gender politics a core issue for Conrad, but Mallios sees a feminist impulse behind his portrayal of masculine struggles without women, and McIntire argues that he "uses race and gender together" to uphold oppressive barriers between Belgium and the Congo (Mallios 2013, 132; McIntire 2002, 261). The non-human natural world is brought into this political discussion by Jeffrey Myers, but only to be cast as another victim of human political oppression: "the Eurocentric bias inscribed in the text is the narrow point of a broader anthropocentric bias" (2001, 98).

Political discussions of novels that are not openly political can in theory be illuminating, but the political interpretations of *Heart of Darkness* have tended to bypass its meaning. The novel depicts an encounter with human moral flexibility in the amoral natural world. By focusing on whether it is complicit in political oppression or opposed to it, scholars discuss its meaning within the comfort of clear-cut human moral divisions. Even when historical allowances are made for Conrad's perspective, his tendencies toward racism and sexism are conceived as transcendentally evil, and his tendencies toward racial and sexual equality are conceived as transcendentally good. Mallios's good Conrad—read through the feminist Mary Austin—portrays differences between people as a path to "larger political goals of solidarity and justice" (2013, 133). McIntire's evil Conrad insists on differences between races and genders to "guard the difference between male and female forms of knowledge" (2002, 280). Tim Christensen calls *Heart of Darkness* "prescient" because, as he claims, it portrays differences between white and black bodies as "a foundational fetish of imperial ideology" (2012, 24). Jesse Oak Taylor agrees that *Heart of Darkness* "productively" exposes masculine imperialist identity, but notes that its racism is "inexcusable" (2012, 208). In this way, politically motivated scholars have placed Conrad with the angels or devils in a perceived historical progression toward justice—and made the type of certain moral judgment about him that Marlow cannot make about Kurtz. Though Conrad might have appreciated the idealist convictions of such scholars, he would no doubt have been as skeptical of their moral absolutes as he was of the late Victorian ones.

Scholars without overt political agendas have also tended to leave out the cosmic level of *Heart of Darkness*. Ian Watt deals with proxies for the challenge of the amoral cosmos: Conrad's satire of Victorian jingoism and progressivist rhetoric (1980, 155–61). That analysis illuminates the novel's cultural context, but it does not address the cosmological challenge underlying the Victorian rhetoric, or Conrad's own response to that challenge. There is a tradition of explaining *Heart of Darkness* through philosophy: Schopenhauer, Buddhism, or even Nietzsche (Alpert 2016; DeMille 1990; Knowles 1994; Madden 1999). Patrick McCarthy and

Linda Dryden have interpreted the novel through the influence of H. G. Wells (McCarthy 1986; Dryden 2013). These explanations are at most partial—like all studies of cultural influence, they struggle to capture Conrad's idiosyncratic cosmic vision. As Najder observes, Conrad was not a close follower of Schopenhauer, who is the most well-established philosophical influence; and as McCarthy and Dryden acknowledge, Conrad and Wells had fundamental differences of outlook (Najder 2011, 254–56; McCarthy 1986, 59–60; Dryden 2013, 230). Scholars seeking explanation beyond cultural influence have looked to non-naturalistic frameworks that cannot encompass Conrad's naturalism. F. R. Leavis sees the novel as a "simple" contrast between the human soul's "corruption" and "innocence," and considers it unnecessarily embroidered with rhetoric (1950, 180). J. Hillis Miller sees it as a "cacophony of dissonant voices" in harmony only with the poststructuralist impossibility of knowledge and communication (1965, 216). Frederick Crews launched an interpretive tradition relying on psychoanalysis—which acknowledges the importance of family connections, somewhat in line with Conrad's moral ideals, but which departs both from Conrad and reality by framing those connections as pathological (Murfin 1992, 106–8).

As an imaginative evocation of Conrad's cosmic vision, *Heart of Darkness* is unrivalled. His letters and essays contain rhetorical high points and intricate cosmic metaphors laced with irony, but they do not sustain those features in an interconnected way for ninety pages. In *Heart of Darkness*, the cosmic symbolism resembles a symphony. The cosmic scope is fully established in the novel's first scene. The framing narrator describes the Thames stretching "like the beginning of an interminable waterway," into a sea haze that merges with the horizon (1992 [1899], 17). The mood on the cruising yawl is "meditative" (18). When the sun reddens and cools, "as if about to go out suddenly," the former sailors find a sense of permanence in the river: "We looked at the venerable stream not in the vivid flush of a short day that comes and departs for ever, but in the august light of abiding memories" (18). Conrad's cold earth is warded off by a reflection on the British explorers, great or mercenary, carried by the river throughout history: "bearing the sword, and often the torch, messengers of the might within the land, bearers of spark from the sacred fire" (19). It is at this point that Marlow interjects. He observes that the city of London was wilderness "the other day," with Roman triremes on the river (20). He declares that British ideals are not an abiding sacred fire but a wildfire in the grass—a "flicker" that may, with the best possible fortune, only last "as long as the old earth keeps rolling" (20). Yet, Marlow hopes that the ideals may last that long, and that they may be granted "unselfish belief" and "sacrifice" (21). In a few pages, the reader's mind has been elevated to a meditation on transient lives in an eternal natural and cultural setting, then wrenched into a recognition that the setting itself is transient—and then, implored to dedicate its own transient existence to a slightly larger transient ideal.

Marlow's story is told within the cosmic frame of the first scene. The story can be regarded as one long illustration of this transient earth with flickering human ideals—and an explanation of how he came to view the earth in that way. Colonialism is the subject of the story, but on the cosmic level at which the novel is pitched, the colonists represent the sword and the torch: self-interested domination and idealism. Marlow is contemptuous of the self-interested domination—the "robbery with violence" he associates with the Romans, and the "weak-eyed devil of rapacious and pitiless folly" that drives the Belgians (21, 31). The idealist impulse of colonialism is the only thing that "redeems" it (21). In Marlow's account, that impulse is no more specific than Conrad's most general idea of faith or hope: the capacity to believe that the world can be made good. Kurtz reveals the fundamental unreliability of that capacity—because, as Conrad saw it, moral emotions do not have cosmic purpose and altruistic enthusiasm is not an adequate guide for behavior. For Marlow too, Kurtz symbolizes the lack of cosmic justice revealed by naturalistic cosmology. His memory of Kurtz's rotten idealism transforms his friends' comfortable historical vision of the Thames into a flux of unpredictability, culminating in the image of the old earth ceasing to roll. In *Heart of Darkness*, as in his exchanges with Graham, Conrad shifts continuously between ethics and the cosmic mechanism.

Marlow's story is impelled by his need for cosmic meaning. The African wilderness gives him a powerful impression of the naturalistic cosmos. He imagines that he is "travelling back to the earliest beginnings of the world," he feels cursed or welcomed by "prehistoric man," and he compares the noise of a stream to "the tearing pace of the launched earth" (31, 49, 51). The impression disorients him. He becomes lost "in another existence," dwarfed on the silent river, unable to decide whether the wilderness is "like evil or truth" (38, 49). Analyzing his response to the natives on shore, he concludes that "the mind of man is capable of anything—because everything is in it, all the past as well as all the future" (52). He has an acute need to find some guiding human ideals.

But his encounter with the colonists is a crescendo of senselessness, inefficiency, and cruelty. He watches a French man-o-war "incomprehensible, firing into a continent"; he stumbles through broken machines among the "objectless blasting" of a stagnant railway construction, and he ends up in a grove where the native workers are "dying slowly" from disease and useless work (29, 30, 32). The first competent man he meets only worries about proper dress and accounts, but even that is a relief for Marlow (33). It is in this context that he learns of Kurtz, "who had come out equipped with moral ideas of some sort" (46). Marlow's increasing desire to meet Kurtz is an increasing need to find human moral ideals—both to guide his own mind through the naturalistic cosmos and to learn the truth about human morality.

The idealism Marlow seeks is weighted toward the same simple moral ideals as Conrad's moral attitude. Throughout his narrative, he admires the British government's "devotion to efficiency," a book of seamanship's "honest concern" for "the right way of going to work," and the starving cannibals' "restraint" in not attacking the European crew (21, 53, 58). In his culminating interview with Kurtz's fiancée, he admires her "mature capacity for fidelity, for belief, for suffering" (91). He is proud of his own discipline as he repairs and steers the steamer, and ashamed of unrestrainedly throwing his shoes overboard when they are filled with the helmsman's blood (43–45, 60–64). These ideals, however vague and general, predispose him toward visions of cosmic purpose that stress stable long-term cooperation. He cannot see domination as a source of meaning, though he does not see it as a transcendental evil either: "strength is just an accident arising from the weakness of others" (21). The Belgian colonists represent a less admirable version of the domination he illustrates with the Romans, and it is taken to its extreme in the Eldorado Exploring Expedition he sees passing through camp:

> Their talk … was the talk of sordid buccaneers: it was reckless without hardihood, greedy without audacity, and cruel without courage; there was not an atom of foresight or of serious intention in the whole batch of them, and they did not seem aware these things are wanted for the work of the world. To tear treasure out of the bowels of the land was their desire, with no more moral purpose at the back of it than there is in burglars breaking into a safe. (46)

Had these adventurers been hardy and courageous, they might have stepped out of *The Lost World*. Daring quests for land and trophies was an adequate cosmic purpose for Arthur Conan Doyle's heroes. For Marlow, who shares Conrad's moral attitude, it is not. He needs not only courage and hardihood (which he grants the Romans), but a "moral purpose at the back of it" that sustains his idea of civilization at its best. He needs to attach his self-discipline to a more general serious intention, backed up by matter-of-fact competence, toward some goal of benefit beyond the individual.

Kurtz's reputation leads Marlow to believe that there is hope of moral ideals among the colonists. He hears that Kurtz wants each colonial station to be "like a beacon on the road toward better things," tasked with "humanising, improving, instructing" (48). Kurtz is described to him as "an emissary of pity, and science, and progress" (40). One of the earliest impressions of this emissary is a sketch in oils he has left behind: a woman walking through darkness, "draped and blindfolded, carrying a lighted torch" (40). Like the initial vision of the eternal Thames, these hints have a dark undercurrent. The colonists speak of Kurtz's idealism in tones of venomous cynicism, and the light on the torchbearer's face is "sinister" (40). It is apparent from the start that Kurtz is a particularly ruthless ivory agent: "he had collected, bartered, swindled, or stolen more ivory than all the other agents together" (63). Nonetheless, Marlow gradually becomes obsessed with meeting Kurtz. He deals with the overwhelming wilderness by focusing on his steamer's progress, and he imagines it going "toward Kurtz—exclusively" (51). When a skirmish with the natives seems to indicate that Kurtz is dead, Marlow's response is plainly existential:

> There was a sense of extreme disappointment, as though I had found out I had been striving after something altogether without substance. I couldn't have been more disgusted if I had travelled all this way for the sole purpose of talking with Mr. Kurtz.... I couldn't have felt more of lonely desolation somehow, had I been robbed of a belief or missed my destiny in life. (63)

In light of the novel's cosmic drama, this moment is not disproportionate. Marlow has become fundamentally disoriented—convinced that the human mind is capable of anything in an inexorable natural world. On the verge of despair, he has invested Kurtz with the guiding purpose that he needs. Conrad anticipates the readers who have not been following his emotional trajectory by having one of Marlow's listeners interrupt the existential disappointment with a "beastly" sigh (63).

At the most general level, Kurtz represents altruistic enthusiasm. Conrad deliberately avoids connecting him to any particular policy, moral system, or practical course of action. What are the better things Kurtz wants? How does he imagine Justice walking through the darkness? What does it mean to humanize, improve, and instruct? The vagueness of the altruism is not in itself a condemnation, either from Conrad or Marlow. But Kurtz's enthusiasm strikes at the heart of Conrad's ambivalence about idealism. Kurtz is famous for his eloquence—his ability to move himself and others with words. This is what Marlow most wants from Kurtz, in his extreme moment of existential despair: "his ability to talk, his words—the gift of expression" (63). He wants to be moved into a sense of moral purpose. For Conrad, the sway of rhetoric was the source of all greatness and upheaval in human society (1996 [1912], xi). But he himself stood back from that sway. His skeptical life strategy was a way to guard against the exact power that Marlow seeks in Kurtz.

Marlow's susceptibility to Kurtz's rhetoric is the subject of acute retrospective self-irony. When he describes encountering Kurtz's written report to the "International Society for the Suppression of Savage Customs," it results in one of the novel's most searing passages:

> 'By the simple exercise of our will we can exert a power for good practically unbounded,' etc. etc. From that point he soared and took me with him. The peroration was magnificent, though difficult to remember, you know. It gave me the notion of an exotic Immensity ruled by an august Benevolence. It made me tingle with enthusiasm. This was the unbounded power of eloquence—of words—of burning noble words. There were no practical hints to interrupt the magic current of phrases, unless a kind of note at the foot of the last page, scrawled evidently much later, in an unsteady hand, may be regarded as the exposition of method. It was very simple, and at the end of that moving appeal to every altruistic sentiment it blazed at you, luminous and terrifying, like a flash of lightning in a serene sky: 'Exterminate all the brutes!' (66)

The "flash of lightning" shows rhetoric's extreme power to deceive and self-deceive, and the inadequacy of altruistic enthusiasm as a moral compass. Kurtz poses as a god and impales the heads of dissenters to satisfy his desire for power, while talking of "love, justice, conduct of life"

(73–75). With his eloquent faith, he has guided the minds of numerous people across Europe and Africa, including, for a time, Marlow's own.

Neither Marlow nor Conrad makes a final judgment of the human capacity for faith. Marlow finds the evidence of Kurtz's imaginative influence—the "unspeakable rites" and "ceremonies" offered to him—more disturbing than the physical evidence of his murders (66, 74). These religious tributes reveal humanity's ability to invest pure self-assertion with meaning. Kurtz "can make himself believe anything," and he finally believes only in his own power (89). Marlow is so disturbed by this reality that his own "belief in mankind" withers, and he descends into a depressive delirium on his return to Europe: "It was not my strength that wanted nursing, it was my imagination that wanted soothing" (83, 88). But instead of judging Kurtz and humanity in terms of cosmic right and wrong, Marlow retreats into a Conradian position of observation. While telling his story, he looks on Kurtz the way he looks on himself, the natives, the other colonial agents, and Kurtz's fiancée—the way he describes looking on the steamer's crew of cannibals: "as you would on any human being, with a curiosity of their impulses, motives, capacities, weaknesses, when brought to the test of an inexorable physical necessity" (57). Kurtz, when brought to the test of unrestrained power, dissolves into moral incoherence. There is "barren darkness" in his heart, and his soul has "gone mad," but his rampant self-assertion is not the result of some transcendental evil (83, 85). For Conrad, Kurtz's seduction by those "primitive emotions"—and his power to seduce—is simply a human contingency that is slightly reduced by social restraints (85). The charismatic enthusiasm Kurtz uses to make people kill for his personal benefit is the same ability he used to inspire his own altruistic hopes and the faith of his noble fiancée. That natural reality is enough to unsettle Marlow's imagination.

Conrad presents Marlow's inconclusive judgment of Kurtz as a deliberate challenge. The other colonists, who interpret the situation through their political intrigues, view Marlow and Kurtz as a coalition—first as "the gang of virtue," then as "the party of 'unsound method'" (41, 85). The two of Kurtz's disciples that Marlow meets, his fiancée and a Russian adventurer in the Congo, both treat Marlow as a fellow disciple (74–75, 92). Marlow himself is drily amused by these assumptions: "I found myself lumped along with Kurtz as a partisan of methods for which the time was not ripe"; "I suppose it did not occur to him that Mr. Kurtz was no idol of mine" (75, 79). But his irony does not amount to a denunciation. Marlow is conscious of his imaginative attachment to Kurtz, which culminates in his moment of existential crisis but remains as a half accidental loyalty. He acknowledges that his aloofness from the other colonists, who are conspiring against Kurtz, has made him "Mr. Kurtz's friend—in a way" (79). Even after seeing the impaled heads around Kurtz's hut, Marlow is so repelled by the other colonists' petty and dishonest cruelty that he turns to Kurtz "for relief—positively for relief" (79). But he does so in reluctant terms that register the limitations of his relief. Kurtz is his "choice of nightmares"; he is conflated with the amoral natural forces that seem a comfort next to the colonists' self-serving pretense: "I turned to the wilderness really, not to Mr. Kurtz" (79). Marlow is more disturbed by Kurtz's degradation than are either the mercenary colonists or the blind disciples, and yet, he is more willing to face that degradation than any of them. His imaginative attachment to Kurtz does not prevent him from seeing the full horror of what Kurtz has become. But his awareness of the horror does not prevent him from remaining attached to the capacity for faith that Kurtz represents. That relationship is impossible to categorize in the terms of alignment used by the colonists and the disciples.

Marlow's double vision of Kurtz is crystallized in his reflections on Kurtz's final moment. Kurtz dies in anguish, whispering his vague condemnation "the horror" (86). What is he condemning? He might, as Marlow suspects, be seeing something of his own psychological "deficiency": "I think the knowledge came to him at last—at the very last" (74). But he might also, as Marlow imagines elsewhere, be "embracing, condemning, loathing all the universe" (91). Both of these interpretations are contained in Marlow's most general view of Kurtz's last words: a "burst of sincerity" (83). Until the last, Kurtz has been discoursing on his "right motives"; he has been indulging childish "images of wealth and fame revolving obsequiously round his unextinguishable gift of noble and lofty expression" (85). Marlow hears him muttering about living "rightly" (86). But in his final moment, the dying man abandons his altruistic rhetoric and fantasies of prestige:

> It was as though a veil had been rent. I saw on that ivory face the expression of sombre pride, of ruthless power, of craven terror—of an intense and hopeless despair. Did he live his life again in every detail of desire, temptation, and surrender during that supreme moment of complete knowledge? He cried in a whisper at some image, at some vision—he cried out twice,

> a cry that was no more than a breath.... 'The horror!' He was a remarkable man. After all, this was the expression of some sort of belief; it had candour, it had conviction, it had a vibrating note of revolt in its whisper, it had the appalling face of a glimpsed truth—the strange commingling of desire and hate. (86–87)

For Marlow, Kurtz's moment of sincerity is at once his most admirable and most disturbing. The only purpose Conrad maintained for the non-ethical universe was to be witnessed sincerely. Self-deception, emotional excess, and inconsistency all threatened the mental coherence that was his last bulwark against cosmic futility. Marlow frames sincerity in similarly existentialist terms: "There is a taint of death, a flavour of mortality in lies—which is exactly what I hate and detest in the world—what I want to forget" (42). Whether Kurtz's last words condemn himself, or the world, or his fleeting vision, Marlow sees them as a response to his actual experiences, without hypocrisy or fantasy: "a judgment upon the adventures of his soul on this earth" (87). Kurtz's last eloquent conviction illuminates his mind as it was. In the eyes of Marlow—and Conrad—that is "a moral victory paid for by innumerable defeats" (88). But the nature of the victory is a full revelation of the defeats. When Kurtz's fiancée declares that Kurtz "died as he lived," Marlow's response implies how he feels about the truth he saw: "'His end,' said I, with dull anger stirring in me, 'was in every way worthy of his life'" (94).

Marlow is a personification of Conrad's idea of cosmic observation. Like Conrad, he sees human motivations flowing through "the sea of inexorable time," toward "that oblivion which is the last word of our common fate" (85, 90). And like Conrad, he makes his last stand for a cosmic purpose through the idea of observation with open-eyed resignation. He watches those around him with both sympathy and ironic detachment. He holds his own mind together with his simple Conradian ideals, but he does not imagine that the universe prefers his efficiency and restraint to Kurtz's moment of unrestrained horror. He praises Kurtz for a "summing-up" of which he claims to be incapable himself, but he follows Kurtz's moment of sincerity with his own laconic conclusion (88):

> Droll thing life is—that mysterious arrangement of merciless logic for a futile purpose. The most you can hope from it is some knowledge of yourself—that comes too late—a crop of unextinguishable regrets. (87)

This is the core of Conrad's outlook: life is governed by amoral natural laws from which you must stand back, in skeptical self-reflection, to pursue integrity for your moral self. Kurtz's self-knowledge—such as it is—comes so late that the regrets are immediately extinguished. Marlow's self-knowledge and regrets have become part of him. He notes that his listeners, who know him as he is now, "see more" in his story than he himself did when he arrived in Africa (43).

Marlow's distance from his younger self reflects his development toward Conrad's naturalistic moral outlook. His listeners know him much as Conrad might have wished them to know him: as an ascetic, curious "wanderer" who watches "the changing immensity of life" (19). In the feverish Congo, that same Marlow attached his hopes to Kurtz's rhetoric, losing his shoes and nerve to the river. He yearned for proof of enduring human ideals and was stricken into mental paralysis when he found flexible morality swayed by primeval emotions. As an older man, he calls his meeting with Kurtz "the culminating point" of his experience (22). His culminating experience revealed the conflict between human ideals and amoral nature. Marlow used the same phrase to describe that conflict as Conrad later would: "cruel and absurd mysteries not fit for a human being to behold" (92). The young Marlow could not satisfy his need for justice—or his commitment to sincerity—by telling Kurtz's fiancée the truth. Partly, he was prevented by an old-fashioned chivalry that dictates she should be protected from life's horrors: "the women are out of it" (64). But it would also have hurt his own unsteadied imagination to disturb her grief with Kurtz's depravity: it "would have been too dark" (94). Like Conrad, Marlow prefers the fiancée's "light of belief and love" to Kurtz's "eternal condemnation" (92). Neither faith nor condemnation is a relief to him personally. Instead, like Conrad, he relieves his mind by expressing the full range of his observer's perspective. He finds coherence in his experience and satisfies his sense of justice about Kurtz ("Hadn't he said he wanted only justice?") by shaping it into an imaginative account for his listeners on the Thames (94).

Heart of Darkness manages its readers in a remarkably effective way. Marlow's interaction with his listeners anticipates and guides responses from the least to the most sympathetic. His first comment on the transience of British civilization does not make an impression on his audience: "It was just like Marlow"—"no one took the trouble to grunt even" (20). The framing narrator invites readers to have patience with the peculiar Marlow—to follow him in finding "the meaning of an episode" beyond its apparent events (20). As the tale progresses, this framing narrator becomes Marlow's ideal listener: "on the watch for the sentence, for the word, that would give me the clue to the faint uneasiness inspired by this narrative" (43). Other listeners sigh and grunt at Marlow's imaginative susceptibility, or growl at him to be civil when his humor becomes too harsh, provoking reflections about the impossibility of communication (63, 52, 50). When the tale comes to an end, the silence is broken only by the practical remark that they have lost "the first of the ebb" (95). But the framing narrator sees a different Thames:

> I raised my head. The offing was barred by a black bank of clouds, and the tranquil waterway leading to the uttermost ends of the earth flowed sombre under an overcast sky—seemed to lead into the heart of an immense darkness. (95)

There is no more eternal river that carries eternal cultural ideals. The framing narrator who began by resting in sacred light ends by tracing the darkness of moral transience across the globe. Whether or not he has found his clue to Marlow's narrative, he has transformed his faint uneasiness into a distinct imaginative impression. And so, despite Marlow's frequent despair of being understood, he has transferred his culminating experience to at least one of his listeners.

At the level of cosmic meaning that motivates the novel, *Heart of Darkness* is not widely accessible. It is Conrad's cosmic testimony, translating a personal nightmare journey into a conscious reckoning with the transience and moral flexibility of the human species. Its deepest drama is Marlow's mind facing the cosmos and achieving Conradian skeptical resignation. To experience that drama, readers need to be willing and able to face the amoral cosmos, and they need to understand—if not sympathize with—Conrad's position as cosmic spectator. Conrad realized that most readers would not. He may not have planned to write an allegory of his cosmic vision—as he told Graham, he started the novel with "definite images" rather than with "abstract notions"—but he had some idea of what he had ended up doing (2015, 113). He told his publisher William Blackwood that the novel comes together in the conversation between Marlow and Kurtz's fiancée: "into one suggestive view of a whole phase of life," "something quite on another plane than an anecdote of a man who went mad in the Centre of Africa" (2015, 152). That scene in the novel marks the culmination of Marlow's encounter with the naturalistic human condition. In his hallucinatory overlap between the real and idealized Kurtz, he watches the human capacity for faith and—like Hudson watching the spider—he simultaneously sees the cruelty and beauty of a natural phenomenon. Conrad expected even Graham to be puzzled by this finale—partly because it did not support Graham's "convictions" (at least through nothing "practically effective") and partly because "the idea" was obscurely expressed: "so wrapped up in secondary notions that You—even You!—may miss it" (2015, 113).

Conrad's self-confessed obscurity raises a question: If the cosmic level of *Heart of Darkness* has been consistently overlooked or translated into different terms by competent commentators, does the cosmic level matter to the novel's meaning? One answer is that Conrad matters to the meaning of *Heart of Darkness*. He was concerned with naturalistic cosmology and its implications for human morality, and his personal way of dealing with those issues corresponds to Marlow's reflections in *Heart of Darkness*. Another answer is that the cosmic interpretation makes more general sense of what scholars have described in other terms: self-discovery, satire of Victorian progressivism, Wellsian alienation, Schopenhauerian pessimism, and deconstructive impossibility of reconciliation. The scholars cited in this chapter widely agree that *Heart of Darkness* is about the nightmarish failure of idealism under pressure of an increasingly complex world. As Moore puts it, "'Heart of Darkness' is frequently invoked as a cultural token signifying the 'horror' at the heart of modern Western civilization" (1996, 223). If the novel is seen as an allegory of facing the amoral cosmos, that horror can be recognized as more universal, even if Marlow's response is personal and idiosyncratic.

Clearly, the cosmic level of *Heart of Darkness* was not necessary for it to become canonical. It is possible to appreciate the novel as a satire of colonialism, or as a masterpiece of style about a traumatic experience in an exotic setting—or as an anecdote of a man who went mad in the center of Africa. Conrad's literary confidant Edward Garnett, for instance, admired Marlow's story and considered it a high-water mark of Conrad's talent, though he saw it simply as a story of white depravity outside the restraint of civilization (Murfin 1992, 99). Such particular cultural or ideological readings cannot, however, make the novel come together as it did in Conrad's comment to Blackwood: "the whole 30,000 words of narrative description" (2015, 152). If Marlow's lie simply hid the ugly side of colonialism, or the personal failings of Kurtz—or the masculine world of action, or the existence of racial others—his rhetoric about the beginnings of time and the flexibility of the human mind would truly be excessive. The cosmic imagery that aligns with Conrad's own cosmological reflections would have to be dismissed as bombast. If Kurtz were simply evil and his fiancée simply good, Leavis might have been right to designate Conrad "a simple soul" (1950, 182). If Marlow had dissolved into incoherence and his listeners responded randomly, Miller might have been right that "his tone and meaning are indeterminate" (1965, 219). But *Heart of Darkness* is a clear depiction of a very particular uncertainty . Conrad's response to Garnett discreetly praised the "brave attempt to grapple with the foggishness of H of D, to explain what I myself have tried to shape blindfold" (2015, 157). This claim of blindness comes off as a Marlowian fib next to his hope that Graham—his sparring partner on questions cosmological and idealistic—would catch the novel's "idea."

7.4 Conclusion

As Matthew Arnold and T. H. Huxley found meaning in the growth of culture and Ernst Haeckel in the beauty of the ecosystem, Conrad found meaning in the cosmic spectacle of humanity. His minimal mythology allowed him to acknowledge natural laws as amoral without dwarfing human morality into nothingness. The nervous, stoic Pole who suspended himself outside systems and emotions produced a life strategy that could accommodate Darwin's non-human perspective on humanity. But it is not surprising that Conrad's perspective as remained largely his own. His cosmic vision is psychologically difficult to access and maintain. It becomes more difficult to access because it is adapted to his unusual personality and cultural experience. People oriented to the world of politics, like Conrad's parents and closest correspondents, might find it more of a wrench than Marlow did to face the primeval emotions of Kurtz. Those who are extraverted and assertive, like H. G. Wells, are not likely to be happy with the purpose of simply bearing witness to the universe. As shown by the inconclusive debates among Conrad scholars, moralists tend to overlook his skepticism and skeptics tend to overlook his morality. Those devoted to a particular moral system are uncomfortable with his dismissal of moral systems, and those devoted to moral relativism are uncomfortable with the simple moral ideas that he applied across cultures.

Conrad's place in the canon is as peculiar as his cosmic vision. There is no doubt that he has been widely influential. As John Galsworthy put it, he was "critically accepted from the very start" and read by the literati with "the bewildered fascination of one opening eyes on a new world" (1925; Moore 1996, 223–24). Toward the end of his life, he became so popularly successful that his landing in New York was under siege for hours—and though his later work was recognized by his first critics as inferior, he did not lose the initial respect of those critics (Conrad 2015, 477; Najder 2011, 554–55). When he died in 1924, Woolf declared that "his reputation of later years was, with one obvious exception, undoubtedly the highest in England" (1984 [1925], 223). The obvious exception, Henry James, summed up Conrad's literary reputation twenty years earlier: "No one has *known*—for intellectual use—the things you know, and you have, as the artist of the whole matter, an authority no one has approached" (quoted in Delbanco 1990, 194). Through his combination of adventurous experience and aesthetic and intellectual subtlety, Conrad found a remarkably varied audience. His influence has reached from English school boys to Polish fugitives of the Gestapo, from the poetry of T. S. Eliot to the garish violence of *Apocalypse Now* (Moore 1996, 223–41, 235). He has been a model for moral action and a subject of hagiographic monographs (Moore 1996, 223–41). But he has also alienated people as different as H. G. Wells and Chinua Achebe, Vladimir Nabokov and E. M. Forster. His reputation declined shortly after his death and was only gradually revived during the middle of the twentieth century (Peters 2006, 121–25). Despite his current secure place in the canon, the criticism cited in this chapter suggests that the significance of his literary contribution—both its value and its most basic meaning—remains under vigorous debate.

By understanding Conrad's worldview as an idiosyncratic response to naturalistic cosmology, it is possible to unite the two main tendencies of Conrad scholarship. Conrad was not just a Romantic, or a Realist, or a Romantic Realist. He was an unusual man with a tumultuous life, caught between cultural allegiances, upholding simple naturalistic ideals in response to the natural world he recognized as amoral. He maintained a skeptically detached, tragic commitment to transient human attachments, occasionally enlivened by absurdist humor. That vantage point illuminates his individual authorship, and also the guiding function of verbal art. Conrad's novels—most prominently *Heart of Darkness*—allow us to inhabit the mental world he created for himself. For those who follow Marlow through the darkness, there may be new clarity in the uneasy sense of cosmic transience and chaos, and a new appreciation for the human attachments embodied by steamboat rivets and light. For those who sigh and grunt at him—wishing he would use fewer adjectives and be more civil—there may be a renewed passion to stave off the darkness using systems of their own. In both cases, the symbolic intermingling of human darkness and light may provide mythic texture for the post-Darwinian world. For Conrad himself, as for Marlow, there seems to have been some relief in the shaping and sharing of his cosmic drama. A dark earth with flickering human ideals is less comforting than the eternal torch of the Thames, but also less vulnerable to critical reflection—and more comforting than an earth that is already cold around the sun.

Conrad's combination of the idiosyncratic and the cosmic is its own challenge. Just as he saw the world through his identity, his readers see him through theirs. H. G. Wells, who was emotionally stable and proudly "eschewed dignity," could not take seriously Conrad's "*persona* of a romantic adventurous un-mercenary intensely artistic European gentleman carrying an exquisite code of unblemished honour through a universe of baseness" (1934, 530). He characteristically declared that such personas could be corrected "with a more expert and scientific educational process" (1934, 529). Virginia Woolf, who released her own imaginative neuroticism in unconventional literary form, thought that the problem was rather too much disciplining. She suggested that "Conrad would have gained both in credit and in popularity if he had written what he had to write without this incessant care for appearances" (1984 [1925], 224). More recent critics, framing their identities in terms of gender and ethnicity, have claimed that Conrad is inaccessible to African readers, or to female readers, and that his canonical status should be adjusted accordingly (Achebe 1977; Murfin 1992, 109; Straus 1987, 124).

Retrospective wishes that Conrad be more like us obscure what he actually was—and what is genuinely universal in his worldview. Though his life strategy was weighted toward his personality and culture, it allowed him to admire the fidelity of fiancées and military commanders, the efficiency of sailors and accountants, the sincerity of Henry James, and the restraint of Congolese cannibals. During Conrad's eclipse, in the middle of the twentieth century, these simple ideals appeared to be a relic of the nineteenth century. T. S. Eliot adapted Conrad for the hollow horror of Kurtz, not the moral reflections of Marlow. Even Woolf, who glimpsed Conrad's combination of darkness and light, dismissed his moral ideals in the manner that prevailed during his eclipse: "in this more crowded and complicated world such terse phrases became less and less appropriate"

(1984 [1925], 229). She claimed that his morality was simplistic because he had spent too much time in the small universe of a ship: "There are no masts in drawing rooms; the typhoon does not test the worth of politicians and businessmen" (229). As it happened, few twentieth-century moral attitudes could have been more deliberately adapted to the crowded and complicated world—or more intricately applied to the new moral complexity—than Conrad's terse phrases.

References

Achebe, Chinua. 1977. "An Image of Africa: Racism in Conrad's *Heart of Darkness*." *Massachusetts Review* 17 (4): 782–94.

Alpert, Avram. 2016. "Empires of Enlightenment: On Illumination and the Politics of Buddhism in *Heart of Darkness*." *Journal of Modern Literature* 40 (2): 1–21.https://doi.org/10.2979/jmodelite.40.2.01.

Bohlmann, Otto. 1991. *Conrad's Existentialism*. London: Macmillan.

Christensen, Tim. 2012. "Joseph Conrad's Racial Idea." *Colloquy: Text Theory Critique* 24: 5–31.

Conrad, John. 2008. *Joseph Conrad, Times Remembered: Ojciec jest tutaj*. Cambridge, UK: Cambridge University Press.

Conrad, Joseph. 1992 [1895]. *Almayer's Folly: A Story of an Eastern River*. Edited by Jacques Berthoud. Oxford: Oxford University Press.

———. 1992 [1899]. "Heart of Darkness." In *Joseph Conrad, Heart of Darkness: A Case Study in Contemporary Criticism*, edited by Ross C. Murfin, 17–95 . Basingstoke: Macmillan.

———. 2004 [1904]. "Henry James: An Appreciation." In *The Cambridge Edition of the Works of Joseph Conrad: Notes on Life and Letters*, edited by J. H. Stape and Andrzej Busza, 15–20. Cambridge University Press.

———. 1974 [1904]. *Nostromo: A Tale of the Seaboard*. Harmondsworth: Penguin Books.

———. 2004 [1905]-a. "Autocracy and War." In *The Cambridge Edition of the Works of Joseph Conrad: Notes on Life and Letters*, edited by J. H. Stape and Andrzej Busza, 71–93. Cambridge University Press.

———. 2004 [1905]-b."Books." In *The Cambridge Edition of the Works of Joseph Conrad: Notes on Life and Letters*, edited by J. H. Stape and Andrzej Busza, 9–14. Cambridge University Press.

———. 1996 [1906 and 1912]. *The Mirror of the Sea and a Personal Record*. Edited by Zdzislaw Najder. Oxford: Oxford University Press.

———. 1911. *Under Western Eyes*. London: Methuen & Co. Ltd.

———. 2004 [1917]. "Turgenev." In *The Cambridge Edition of the Works of Joseph Conrad: Notes on Life and Letters*, edited by J. H. Stape and Andrzej Busza, 40–42. Cambridge University Press.

———. 2004. *The Cambridge Edition of the Works of Joseph Conrad: Notes on Life and Letters*. Edited by J. H. Stape and Andrzej Busza. Cambridge University Press.

———. 2010. *Last Essays*. Edited by Harold Ray Stevens. Cambridge: Cambridge University Press.

———. 2015. *The Selected Letters of Joseph Conrad*. Edited by Laurence Davies. Cambridge, New York: Cambridge University Press.

Cox, Charles Brian. 1974. *Joseph Conrad: The Modern Imagination*. London: J. M. Dent and Sons.

Delbanco, Nicholas. 1990. *Group Portrait: Joseph Conrad, Stephen Crane, Ford Madox Ford, Henry James, and H. G. Wells*. New York: Carroll & Graf.

DeMille, Barbara. 1990. "Cruel Illusions: Nietzsche, Conrad, Hardy, and the 'Shadowy Ideal.'" *Studies in English Literature, 1500–1900* 30 (4): 697–714. https://doi.org/10.2307/450567.

Dryden, Linda. 2013. "'The Difference between Us': Conrad, Wells, and the English Novel." *Studies in the Novel* 45 (2): 214–33.https://doi.org/10.1353/sdn.2013.0027.

Erdinast-Vulcan, Dapha. 1999. *The Strange Short Fiction of Joseph Conrad: Writing, Culture, and Subjectivity*. Oxford: Oxford University Press.

Fyfe, Herbert C. 1900. "How Will the World End?" *Pearson's Magazine:* 86–94.

Galsworthy, John. 1925. "Reminiscences of Conrad." *Scribner:* 3–9.

Griffith, John Wylie. 1995. *Joseph Conrad and the Anthropological Dilemma: "Bewildered Traveller."* Oxford, Clarendon, New York: Oxford University Press.

Hampson, Robert. 1985. "Determining Fictions." *English* 34 (150): 259–65. https://doi.org/10.1093/english/34.150.259.

———. 2012. *Conrad's Secrets*. Basingstoke, Hampshire, New York: Palgrave Macmillan.

Harpham, Geoffrey Galt. 1996. *One of Us: The Mastery of Joseph Conrad*. Chicago, London: The University of Chicago Press.

Hunter, Allan. 1985. *Joseph Conrad and the Ethics of Darwinism: The Challenges of Science*. London, Sydney: Croom Helm.

Jameson, Frederic. 1982. *The Political Unconscious: Narrative as a Socially Symbolic Act*. Ithaca, NY: Cornell University Press.

Knowles, Owen. 1994. "'Who's Afraid of Arthur Schopenhauer?': A New Context for Conrad's Heart of Darkness." *Nineteenth-Century Literature* 49 (1): 75–106. https://doi.org/10.2307/2934045.

Leavis, F. R. 1950. *The Great Tradition: George Eliot, Henry James, Joseph Conrad*. New York: George W. Stewart.

Levine, George. 1991. *Darwin and the Novelists: Patterns of Science in Victorian Fiction*. Chicago, IL: University of Chicago Press.

Madden, Fred. 1999. "The Ethical Dimension of Heart of Darkness and Lord Jim: Conrad's Debt to Schopenhauer." *Conradiana* 31 (1): 42–62.

Mallios, Peter Lancelot. 2013. "Democracy and Difference: Mary Austin, Joseph Conrad, and Global Feminism." *Studia Neophilologica* 85 (1): 125–35. https://doi.org/10.1080/00393274.2012.751668.

Martin, Regina. 2015. "Absentee Capitalism and the Politics of Conrad's Imperial Novels." *PMLA* 130 (3): 584–98. https://doi.org/10.1632/pmla.2015.130.3.584.

McCarthy, Patrick A. 1986. " *Heart of Darkness* and the Early Novels of H. G. Wells: Evolution, Anarchy, Entropy." *Journal of Modern Literature* 13 (1): 37–60.

McIntire, Gabrielle. 2002. "The Women Do Not Travel: Gender, Difference, and Incommensurability in Conrad's *Heart of darkness*." *Modern Fiction Studies* 48 (2): 257–84. https://doi.org/10.1353/mfs.2002.0032.

Mencken, H. L. 1917. *A Book of Prefaces*. New York: A.A. Knopf.

Miller, David. 1990. *W. H. Hudson and the Elusive Paradise*. London: Macmillan.

Miller, Joseph Hillis. 1965. *Poets of Reality: Six Twentieth-Century Writers*. Cambridge: The Belknap Press of Harvard University Press.

Moore, Gene M. 1996. "Conrad's Influence." In *The Cambridge Companion to Joseph Conrad*, edited by J. H. Stape, 223–41. Cambridge: Cambridge University Press.

Murfin, Ross C. 1992. "A Critical History of *Heart of Darkness*." In *Joseph Conrad, Heart of Darkness: A Case Study in Contemporary Criticism*, edited by Ross C. Murfin, 99–114 . Basingstoke: Macmillan.

Myers, Jeffrey. 2001. "The Anxiety of Confluence: Evolution, Ecology, and Imperialism in Conrad's *Heart of Darkness*." *Interdisciplinary Studies in Literature and Environment* 8 (2): 97–108. https://doi.org/10.1093/isle/8.2.97.

Najder, Zdzislaw. 2011. *Joseph Conrad: A Life*. Rochester: Camden House.

O'Hanlon, Redmond. 1984. *Joseph Conrad and Charles Darwin: The Influence of Scientific Thought on Conrad's Fiction*. Edinburgh: Salamander Press.

Peters, John G. 2006. *The Cambridge Introduction to Joseph Conrad*. Cambridge, UK: Cambridge University Press.

Renner, Stanley. 1975. "The Garden of Civilization: Conrad, Huxley, and the Ethics of Evolution." *Conradiana* 7 (2): 109–20.

Roussel, Royal. 1971. *The Metaphysics of Darkness: A Study in the Unity and Development of Conrad's Fiction*. Baltimore: Johns Hopkins Press.

Straus, Nina Pelikan. 1987. "The Exclusion of the Intended from Secret Sharing in Conrad's *Heart of Darkness*." *Novel: A Forum on Fiction* 20 (2): 123–37. https://doi.org/10.2307/1345873.

Taylor, Jesse Oak. 2012. "White Skin, White Masks: Joseph Conrad and the Face(s) of Imperial Manhood." *Conradiana* 44 (2–3): 191–210. https://doi.org/10.1353/cnd.2014.0009.

Thorburn, David. 1974. *Conrad's Romanticism*. New Haven: Yale University Press.

Tsao, Tiffany. 2008. "Conrad and Exploratory Science." *Conradian* 33 (1): 43–56.

Watt, Ian. 1980. *Conrad in the Nineteenth Century*. London: Chatto & Windus.

Watts, Cedric. 1996. "Heart of Darkness." In *The Cambridge Companion to Joseph Conrad*, edited by J. H. Stape, 45–62. Cambridge: Cambridge University Press.

Wells, H. G. 2005 [1895]. *The Time Machine*. Edited by Patrick Parrinder. London: Penguin Books.

———. 1934. *Experiment in Autobiography: Discoveries and Conclusions of a Very Ordinary Brain (Since 1866)*. New York: The Macmillan Company.

Williamson, Karina. 1985. " *Joseph Conrad and the Ethics of Darwinism*. By Allan Hunter." *Notes and Queries* 32 (3): 406–7. https://doi.org/10.1093/notesj/32.3.406.

Woolf, Virginia. 1984 [1925]. *The Common Reader*. New York: Harcourt.

CHAPTER 8

The Unimaginable Place in Nature

We are a way for the Cosmos to know itself.
—Carl Sagan, *Cosmos*.

Science carries us into zones of speculation, where there is no habitable city for the mind of man.
—Robert Louis Stevenson, "Pulvis et Umbra"

How far is it possible to avoid the evolutionary myths I have described? Could we see our place in nature precisely and consistently as it is? Well, try to imagine that. Picture a human mind capable of processing human evolution with perfect accuracy and equipoise: a mind that does not personify abstractions, project human values onto nature, struggle with the geological time scale, or baulk at global extinction rates. That thought experiment is about as difficult as the one demanded by H. G. Wells in "The Man of the Year Million" (1893). Wells told his Victorian readers that they would develop into tentacled heads who fed through suction mouths. Such a physical existence is no more alien to a human mind than an existence without biases and passions—in fact, Wells' promenading heads may be easier to imagine (ask readers of *War of the Worlds*). How would humans with fully objective minds think and feel? Would they have stories of humanity's beginnings, individual self-narratives, ideas about good and evil? They are as hard to imagine in their everyday activities as

E. Jonsson, *The Early Evolutionary Imagination*, Cognitive Studies in Literature and Performance, https://doi.org/10.1007/978-3-030-82738-0_8

Edward Bulwer-Lytton's Vril-ya or Robert Louis Stevenson's Hyde. Of course, that does not mean they won't develop one day—through genetic changes, cybernetic intervention, or both. But if they do, they will hardly be recognizable as human.

The scientists and literary authors in this study did not possess fully objective minds. Modern studies of cognitive biases suggest that 160 years have not produced radical psychological transformations in the human species. For the present, most of us contend with the same limitations as the Victorians—the same tendencies to see nature and culture as opposing forces, to form past and future into narrative structures featuring human characters, and to ascribe human meaning to the cosmos. These limitations do not condemn us to live entirely in our imaginations. Evolutionary theory has, after all, successfully guided biological research for most of the twentieth century and beyond. How far we embrace or avoid myths about evolution will probably remain individually variable—at least until the age of promenading heads. But most of us will be poised in some balancing act between difficult truths and soothing fictions. Such balancing acts are visible among Victorian thinkers, from Huxley's pickled biped to the various glorious purposes based on cultural growth, political systems, conquests of nature, or unity with the ecosystem. In literature, the same balancing acts produced imaginative self-images: Conrad's cosmic observer, London's socialist superman, Wells' Mind of the Race, Stevenson's dual man, Edward Bulwer-Lytton's menacing angels, and the Missin'-Link conquerors of Arthur Conan Doyle. Who are we? As long as we resemble what we presently are, that question is likely to follow us. The imaginative answers of the past are likely to echo alongside the current scientific anatomy.

Seeing the world objectively is not a delusory goal or a simple perspectival shift. It is a difficult achievement that has been slowly approximated throughout scientific history. No single consciousness contains the full extent of our accumulated knowledge. In that respect, it might seem quixotic for any single person to explain naturalistic myth-making. Darwinian evolution is an imaginative problem that is essentially insoluble—a vision of nature and human nature so hard to sustain that it taxed the minds of people like Charles Darwin, T. H. Huxley, and Joseph Conrad. Can anyone understand evolution well enough to explain their difficulties from an evolutionary perspective? Yes, I believe we can—by drawing on research and reflection from over one and a half centuries. Available to us is scientific knowledge to which the Victorians did not have access—and also the insights of the Victorian authors themselves. Though no author in this study entirely avoids transforming evolution imaginatively, none of them misconstrues it so that it becomes unrecognizable. Even Stevenson, who saw Darwinian nature as a crucible for the Christian soul, had a conception of Darwinian nature. He felt the moral weight of the struggle for existence with an intensity that is instructive for the secular mind. London and Conrad both questioned the wisdom of uncovering painful cosmic truths, and they both felt the vulnerability of human consciousness firsthand through depression. But they both faced those cosmic truths with remarkable directness. Their courage is fortifying. By analyzing these early myth-makers of evolution, I have not aimed to solve the imaginative problem with which they struggled. I have aimed to clarify and explain the problem, and to show how literature helps people live with it.

The most powerful imaginative challenges of evolution remain today what they were in the late nineteenth century. Despite the aid of computers, geological time and minute statistical variation still strain the human mind. Details about social evolution and gene-culture coevolution do not eliminate the moral relativism with which Conrad struggled, or the need for large-scale meaning that led Ernst Haeckel to equate his monistic world with God. How are these challenges met today? Intellectual debates about the human condition tend to cordon off culture from nature as if the two were different realms. In the 1990s, biologist Stephen Jay Gould famously declared that facts and values are "non overlapping magisteria" (1999, 49–90). Those "two magisteria" roughly correspond to the "two cultures" of nature and culture that were vigorously debated during the second half of the twentieth century: one culture of physical and natural science, one of humanistic learning, moral values, and artistic appreciation (Snow 1998 [1959]; Pinker 2003; Wilson 1975). According to a recent survey, this basic split remains in the minds of academics who study human behavior (Carroll et al. 2017). Despite an almost universal agreement that behavior is produced by an interaction of genes and culture, one group of academics consistently emphasizes culture while another group consistently emphasizes genes. Those who emphasize culture believe that culture develops independently of biology. Several of these academics declare outright that they envision nature and culture as separate worlds: culture produces identity and mental life while genes shape bodies; human experience, values, and artistic products are inaccessible to science (25–26).

In modern popular science, nature and culture tend to become symbols of good or evil. The two ancient narratives persist: culture has raised us happily above nature's cruelty or separated us sadly from its virtue. Primatologist Franz de Waal exemplifies the latter narrative in *The Age of Empathy: Nature's Lessons for a Kinder Society* (2009). He wants to save nature from its cruel reputation by highlighting natural "empathy and social connectedness" (2009, 5). That ambition is reflected in the titles of his other works: *Peacemaking Among Primates* (1989), *Good Natured* (1996), *Primates and Philosophers* (2006), *Mama's Last Hug* (2019). In de Waal's cosmic narrative, our cultural separation from nature becomes a misfortune similar to the fall from Eden—a vanity to be diminished, as in the title of one of his recent books: *Are We Smart Enough to Know How Smart Animals Are?* (2016). The narrative against which he defines himself can be exemplified by Richard Dawkins's *The Selfish Gene*, first published in 1976. Dawkins declares baldly that we should "*teach* generosity and altruism, because we are born selfish" (1989, 3). In this cosmic narrative, nature is a realm of vicious impulses from which only human culture offers reprieve: "Let us understand what our own selfish genes are up to, because we may then at least have the chance to upset their designs, something that no other species has ever aspired to" (3). Though Dawkins and de Waal profess the same naturalistic world view, their opposite ways of dealing with amoral nature greatly affect their emphases. Culture is our dangerous vanity to one and our sole strand of hope to the other. Popular science remains strongly swayed by these two narrative tendencies. Some books focus on the natural origins of morality, such as anthropologist Christopher Boehm's *Moral Origins* (2012), psychologist Jonathan Haidt's *The Righteous Mind* (2012), and many works by biologist David Sloan Wilson: *Unto Others* (1998), *Darwin's Cathedral* (2002), *The Neighborhood Project* (2011), and *Atlas Hugged* (2020). Others celebrate the power of culture to direct human biology, such as psycholinguist Steven Pinker's *The Better Angels of Our Nature* (2011), anthropologist Joseph Henrich's *The Secret of Our Success* (2015), and the tellingly named bestsellers of historian Yuval Noah Harari: *Sapiens* (2014) and *Homo Deus* (2016). Books in both traditions describe real aspects of the human condition—and, to varying degrees, they all acknowledge its complex interactions of genes, culture, individuals, and groups. Nonetheless, their emphasis on either nature or culture as a source of moral values appeals to the human imaginative need to envision great clashing forces of good and evil.

Popular science that evokes our littleness on the cosmic scale still concludes by implying cosmic purposes for humanity. Even Dawkins's *Selfish Gene* elevates its argument for biological self-knowledge to the cosmic: "Intelligent life on a planet comes of age when it first works out the reason for its own existence"—that is, when it has "discovered evolution" (1989, 1). We seem small in Steven Weinberg's account of the universe's *First Three Minutes* (first published in 1977), in Stephen Hawking's *Brief History of Time* (first published in 1988), in Carl Sagan's meditation on Earth as a *Pale Blue Dot* (1994), and in Geoffrey West's aptly named quest for the laws of life, ecosystems, and cities: *Scale* (2017). But in each case, we have inherent duties: to steward the Earth, to appreciate the beauty of the universe, and—perhaps most importantly—to explore it through science. West frames his own search for unifying systems as "a defining characteristic of *Homo sapiens*, manifested in our multitudinous creeds, religions, and mythologies" (2017, 427). Weinberg's *First Three Minutes* ends by identifying science as "one of the very few things that lifts human life a little above the level of farce, and gives it some of the grace of tragedy" (1993, 155). Hawking claims to act on "humanity's deepest desire for knowledge": "ever since the dawn of civilization, people have not been content to see events as unconnected and inexplicable" (1998, 14). Sagan's *Cosmos* transitions with particular exuberance from humanity's insignificance to its purpose:

> The size and age of the Cosmos are beyond ordinary human understanding. Lost somewhere between immensity and eternity is our tiny planetary home. In a cosmic perspective, most human concerns seem insignificant, even petty. And yet our species is young and curious and brave and shows much promise. In the last few millennia we have made the most astonishing and unexpected discoveries about the Cosmos and our place within it, explorations that are exhilarating to consider. They remind us that humans have evolved to wonder, that understanding is a joy, that knowledge is prerequisite to survival. I believe our future depends on how well we know this Cosmos in which we float like a mote of dust in the morning sky. (1980, 1–2)

Sagan personifies our species as a young scientist—perhaps raised in the Space Age—compelled by nature to understand nature. In the TV series produced alongside the book, he draws out the full implication of the same passage: "The Cosmos is also within us. We're made of star stuff. We are a way for the Cosmos to know itself" (Sagan et al. 1980, episode

1, 5:59). The human mind, first dwarfed into nothingness on its planetary home, returns as the consciousness of the universe—a vessel that contains the vastness that dwarfed it. This image combines Huxley's celebration of cumulative culture with Haeckel's celebration of nature. Humanity is connected to an inherently meaningful natural world, and we are also singled out as the mind of that natural world.

Clearly, scholars and scientists are still finding substitutes for what the Victorians lost. Physicist Sean Carroll faces this problem directly in *The Big Picture: On the Origin of Life, Meaning, and the Universe Itself* (2016). He addresses the modern need for "existential therapy" in a way that would have resonated with Darwin's contemporaries (2016, 3):

> Finally . . . we confront the hardest problem of all, that of how to construct meaning and values in a cosmos without transcendent purpose. A common charge against naturalism is that such a task is simply impossible: without something beyond the physical world to guide us, there is no reason to live at all, and certainly no reason to live one way rather than another. Some naturalists respond by agreeing, and getting on with their lives; others react strongly the other way, by arguing that values can be determined scientifically just as much as the age of the universe can be. (5)

Carroll occupies neither of these positions. His solution, which he calls "poetic naturalism," is in line with the most practical version of Conrad's outlook: "accepting that values are human constructs, but denying that they are therefore illusory or meaningless" (5). Like many Victorians (and unlike Conrad), Carroll is optimistic about his own answer. He believes that humanity can avoid the abyss of moral relativism by building "conceptual jet packs" (10). However, he also acknowledges how hard it has proved to assimilate naturalistic cosmology: "the dominant image of the world remains one in which human life is cosmically special and significant, something more than matter in motion" (14). This claim would be borne out by the simple fact that most humans remain religious (Zuckerman 2007). But Carroll is focusing on the secular societies that formally accept naturalistic cosmology: "many continue on as if nothing has changed; others revel in the new order, but placidly believe that adjusting our perspective is just a matter of replacing a few old homilies with a few new ones" (2016, 10). That claim is borne out by a glance at modern popular science and research on human behavior.

Science as a cosmic duty or a saving grace, human wonder as nature's self-regard, genes and culture as Jekyll and Hyde. Many of the new homilies are over a hundred years old. Some come through direct influence. One of de Waal's early books pays homage to Petr Kropotkin, the nineteenth-century anarcho-communist who expected unfettered Darwinian nature to produce harmonious cooperation (1996, 21–23). Dawkins aligns himself with the Victorians who emphasized competition, and echoes Tennyson's lament: "I think 'nature red in tooth and claw' sums up our modern understanding of natural selection admirably" (1989, 2). In *Dreams of a Final Theory*, Weinberg updates Huxley's deep-time lecture on a piece of chalk (1994, 20–33). Victorian mythologies seem to haunt modern moral foundations, from *Might is Right* supremacism and Spencerian-Nietzschean individualism to Darwin's expanding circles of sympathy and the Bio-Optimists reviewed by Wells. UNESCO and the UN, organizations that provide international moral guidelines, were cultivated by the son and the disciple of Darwin's bulldog (Huxley 1946; Wells 1934, 592–624). But the most significant influence is the scientific worldview to which the Victorians responded in the first place. Behind every secular ideology or personal system of meaning is some idea of the human condition within the naturalistic cosmos. And people contend with naturalistic cosmology according to the patterns of human psychology. Are human beings at their best when left to their natural impulses or when disciplined against them? Should we look to scientists to guide society, or are values outside the scope of science? These questions—and their various passionate answers—still soothe and direct some of our greatest minds.

A study of Darwinism in modern fiction would likely reveal fascinating variations on the old patterns. Human self-images are now complicated by genetics, climate change, artificial intelligence, and the prospect of extraterrestrial colonies. These new subjects are explored in new media, but they are tied into the old concerns. In the TV series *Orphan Black* (2013–2017), genetically modified clones fight Neolution, an evil company hailing from Victorian eugenicists. In the animated movie *WALL-E* (2008), an endearingly tubby humanity leaves its orbiting space crafts and begins re-adapting to the rejuvenating Earth. Dystopian movies routinely depict ecological and biological disasters. In the broodingly stylized *Gattaca* (1997), the government chokes society by controlling human genetics; in the rambunctiously satirical *Idiocracy* (2006), human intelligence has been destroyed because evolution has been left to chance. The dark thriller series *Fortitude* (2015–2017) explores human psychology by using the ichneumon wasps that made Darwin question benevolent creation. Lars von Trier's *Melancholia* (2011) portrays the calm of depressive people as the Earth is destroyed by a rogue planet, and *Europa Report* (2013) documents the tragic end of a mission to find extraterrestrial life. These new forms of fiction clearly help us cope with our naturalistic self-image in somewhat the same way as the Victorian forms did. Humanity is still presented as heroic, comic, or stoic in its meeting with the non-human cosmos. There are still adventure stories, horror scenarios, children's tales, and political allegories. The modern myths, too, range from simple entertainment to psychologically sophisticated existential meditation.

Modern literature has engaged human evolution across a range of genres and styles. Among realist prehistoric fiction, one can find both Jean M. Auel's romance erotica *Earth's Children* (1980–2011) and William Golding's fine Neanderthal tragedy *The Inheritors* (1955) (Carroll 2004; Ruddick 2009). The twentieth century produced a long line of classic science fiction novels that place us in nature, from Aldous Huxley's *Brave New World* (1932) and Arthur C. Clarke's *2001: A Space Odyssey* (1968) to more recent ecological epics and political allegories—Ursula Le Guin's *The Word for World is Forest* (1976), Margaret Atwood's *The Handmaid's Tale* (1985), Brian Aldiss's *Helliconia Trilogy* (1982–1985), and Kim Stanley Robinson's *Mars Trilogy* (1993–1996). Some of these science fiction novels look like direct counters to the heroic narratives of popular science. Kurt Vonnegut's *Galapagos*, first published in 1985, drily wishes that humanity would evolve into seals: "People's brains were much too big and untruthful to be practical" (1999, 207). Douglas Adams's "trilogy of five books," *Hitchhiker's Guide to the Galaxy*, revels in surrealism because a clear view of the universe (through a device called the Total Perspective Vortex) will destroy your brain: "If life is going to exist in a Universe of this size, then the one thing it cannot afford is a sense of proportion" (2005 [1980], 77). In Adams's overwhelming universe, humanity's cosmic purpose is dispatched with famous cheek: "I would like to know … the Ultimate Question of Life, the Universe and Everything. All we know about it is that the Answer is Forty-two, which is a little aggravating" (1982, 226).

There is a world of twentieth-century naturalist literature open for investigation. Golding's *Lord of the Flies* (1954) is an allegory of human nature published within a year of his meditation on the Neanderthals.

John Steinbeck, Golding's fellow Nobel laureate, built his worlds on ecology and pragmatist philosophy (Browne 2007, 51). In *The Grapes of Wrath*, he navigates the relationship between science, humans, and the rest of nature:

> Carbon is not a man, nor salt nor water nor calcium. He is all these, but he is much more; and the land is so much more than its analysis. The man who is more than his chemistry, walking on the earth, turning his plow point for a stone, dropping his handles to slide over an outcropping, kneeling in the earth to eat his lunch; that man who is more than his elements knows the land that is more than its analysis. But the machine man, driving a dead tractor on land he does not know and love, understands only chemistry; and he is contemptuous of the land and of himself. (2002 [1939], 115–116)

In Steinbeck's vision, humans who are separated from nature become meaningless mechanisms. But what makes us "more than our chemistry"? It is the cumulative culture that produced both chemistry and machines: "For man, unlike any other thing organic or inorganic in the universe, grows beyond his work, walks up the stairs of his concepts, emerges ahead of his accomplishments" (150). The puzzles of naturalism remain equally complex toward the end of the century. Ian McEwan's *Enduring Love* (1997) centers on a science journalist whose worldview, informed by evolutionary psychology, is challenged when a stranger becomes obsessed with him. Jim Crace's *Being Dead* (1999) chronicles the decomposition of a murdered couple while meditating on the value of their finite lives: "The doctors of zoology were out of time, perhaps, but they can be rescued from the dunes by memory, receding, and tucked up in their waking beds again, still tenants of the room" (8). These modern naturalists may both illuminate and be illuminated by the universal imaginative challenges they face. Though some have no doubt exchanged old homilies for new ones, others have faced the ancient question "What is a man?" in its dauntingly more specific version: "What is a human animal, following the urges and traditions of survival through an expanse of recombining matter?" Is it a blind buffoon, a tragic warrior, a mechanism or a systematizer, or a dance of self-aware star stuff? Can it see its place in nature?

The book you are currently reading is not exempt from evolutionary myth-making. Analyzing human myth-making can itself be a way to "imagine a place in nature": a way to find human meaning in the amoral natural world. Is our existence made more meaningful if we understand how and why we project human meaning onto the cosmos? From a non-human perspective, of course, the answer is no. But personal values easily take on a cosmic hue when rhetoric runs high. In a 2009 interview, evolutionary literary theorist Joseph Carroll mirrors Sagan's movement from cosmic insignificance to cosmic quest. Carroll paints humanity as "a tiny blip in the most recent moments in the almost unimaginable trajectory of deep time," then as "special" because "we are able to look back over deep time and recognize our own place in it" (2011, 58). In his eyes, humanity's quest is not so much understanding planets and stars as it is understanding "the human imagination": "the means through which we know everything else" (2011, 58). Scientist and literary scholar: each emphasizes his own subject within a larger quest for knowledge, which is rhetorically colored with the cosmos. The book in your hands—with its case for understanding imaginative culture scientifically—is rhetorically allied to both these modern myths.

One lesson from evolutionary history is that life is full of ineradicable challenges that allow degrees of success. Imagining the unimaginable is not unlike staving off daily hunger and thirst, or living with the awareness of death. Through cumulative efforts, humans have built an extraordinarily complex predictive model of the world—including some idea of what kindles and corrupts the human imagination. Literature and the other arts have accompanied these efforts. Metaphors and stories have shaped the frames within which we understand unimaginable facts. Since Darwin's day, there have been new mythological constructs that soothe the human mind into complacency, but there have been others that expand the human mind across deep time and outer space. The balance remains delicate. According to Sean Carroll, "the ground has disappeared beneath us, and we are just beginning to work up the courage to look down" (2016, 10). That "just" is fair vocabulary on a physicist's time scale, but it would have been equally true one and a half centuries ago. And on the human time scale, generations matter. Modern humans may form their own imaginative worldviews in darkness that is peopled with ape men and evolved angels, grasped at by Jekylls and Hydes, studied by the coldly smiling Wolf Larsen, burned by the dying light of Eloi and Morlocks, and pitied by the haunted eyes of Marlow. We may learn from the terror, denial, humor, and courage of those who looked down before us.

References

Adams, Douglas. 1982. *Life, the Universe and Everything*. New York: Del Rey/Ballantine Books.

———. 2005 [1980]. *The Restaurant at the End of the Universe*. New York: Del Rey Books.

Atwood, Margaret. 1985. *The Handmaid's Tale*. Toronto: Seal Books.

Boehm, Christopher. 2012. *Moral Origins: The Evolution of Virtue, Altruism, and Shame*. New York: Basic Books.

Browne, Neil W. 2007. *The World in Which We Occur: John Dewey, Pragmatist Ecology, and American Ecological Writing in the Twentieth Century*. Tuscaloosa: University of Alabama Press.

Carroll, Joseph. 2004. "Adaptationist Criteria of Literary Value: Assessing Kurtén's *Dance of the Tiger*, Auel's *The Clan of the Cave Bear*, and Golding's *The Inheritors*." In *Literary Darwinism: Evolution, Human Nature, and Literature*, edited by Joseph Carroll, 163–186. New York; London: Routledge.

———. 2011. *Reading Human Nature: Litereary Darwinism in Theory of Practice*. Albany: State University of New York Press.

Carroll, Joseph, John A. Johnson, Catherine Salmon, Jens Kjeldgaard-Christiansen, Mathias Clasen, and Emelie Jonsson. 2017. "A Cross-Disciplinary Survey of Beliefs about Human Nature, Culture, and Science. *Evolutionary Studies in Imaginative Culture* 1 (1): 1–32. https://doi.org/10.26613/esic/1.1.2.

Carroll, Sean M. 2016. *The Big Picture: On the Origin of Life, Meaning, and the Universe Itself*. New York: Penguin Random House.

Clarke, Arthur C. 1968. *2001: A Space Odyssey*. New York: New American Library.

Crace, Jim. 1999. *Being Dead*. New York: Picador USA.

Dawkins, Richard. 1989. *The Selfish Gene*. Oxford: Oxford University Press.

de Waal, Frans. 1989. *Peacemaking Among Primates*. Cambridge: Harvard University Press.

———. 1996. *Good Natured: The Origins of Right and Wrong in Humans and Other Animals*. Cambridge: Harvard University Press.

———. 2006. *Primates and Philosophers: How Morality Evolved*. Princeton: Princeton University Press.

———. 2009. *The Age of Empathy: Nature's Lessons for a Kinder Society*. New York: Three Rivers Press.

———. 2016. *Are We Smart Enough to Know How Smart Animals Are?* New York: W. W. Norton.

———. 2019. *Mama's Last Hug: Animal Emotions and What They Tell Us About Ourselves*. New York; London: Norton & Company.

Golding, William. 1954. *Lord of the Flies: A Novel*. London: Faber and Faber.

———. 1955. *The Inheritors*. London: Faber and Faber.
Gould, Steven J. 1999. *Rocks of the Ages: Science and Religion in the Fullness of Life*. New York: Ballantine Pub. Group.
Haidt, Jonathan. 2012. *The Righteous Mind: Why Good People Are Divided by Politics and Religion*. New York: Pantheon Books.
Harari, Yuval Noah. 2014. *Sapiens: A Brief History of Humankind*. New York: Harper.
———. 2016. *Homo Deus: A Brief History of Tomorrow*. New York: Harper.
Hawking, Stephen. 1998. *A Brief History of Time*. New York: Bantam Books.
Henrich, Joseph. 2015. *The Secret of Our Success: How Culture Is Driving Human Evolution, Domesticating Our Species, and Making Us Smarter*. Princeton: Princeton University Press.
Huxley, Aldous. 1932. *Brave New World*. New York: Bantam Classics.
Huxley, Julian. 1946. "UNESCO: Its Purpose and Its Philosophy." *Preparatory Commission of the United Nations Educational, Scientific and Cultural Organization*, 15 June.
Le Guin, Ursula K. 1976. *The Word for World is Forest*. Berkley: Berkley Books.
McEwan, Ian. 1997. *Enduring Love*. London: Vintage Books.
Pinker, Steven. 2003. *The Blank Slate: The Modern Denial of Human Nature*. New York: Penguin Books.
———. 2011. *The Better Angeles of Our Nature: Why Violence Has Declined*. New York: Penguin Books.
Ruddick, Nicholas. 2009. *The Fire in the Stone: Prehistoric Fiction from Charles Darwin to Jean M. Auel*. Middletown, CT: Wesleyan University Press.
Sagan, Carl. 1980. *Cosmos*. New York: Ballantine Books.
———. 1994. *Pale Blue Dot: A Vision of the Human Future in Space*. New York: Random House.
Sagan, Carl, Ann Druyan, and Steven Soter. 1980. *Cosmos*. New York: Cosmos Studios, Inc. Carl Sagan Productions. DVD video.
Sloan Wilson, David. 2002. *Darwin's Cathedral: Evolution, Religion, and the Nature of Society*. Chicago: University of Chicago Press.
———. 2011. *The Neighborhood Project: Using Evolution to Improve My City, One Block at a Time*. New York: Little, Brown and Co.
———. 2020. *Atlas Hugged: The Autobiography of John Galt II*. Redwood Publishing, LLC.
Sloan Wilson, David, and Elliot Sober. 1998. *Unto Others: The Evolution and Psychology of Unselfish Behavior*. Cambridge; London: Harvard University Press.
Snow, C.P. 1998 [1959]. *The Two Cultures*. Cambridge: Cambridge University Press.

Steinbeck, John. 2002 [1939]. *The Grapes of Wrath*. New York: Penguin Books.
Trier, Lars von. 2011. *Melancholia*. United States: Magnolia Pictures.
Vonnegut, Kurt. 1999. *Galapagos*. New York: Dell Publishing.
Weinberg, Steven. 1994. *Dreams of a Final Theory*. New York: Random House.
———. 1993. *The First Three Minutes: A Modern View of the Origin of the Universe*. New York: Basic Books.
Wells, H.G. 1893. "The Man of the Year Million." *Pall Mall Budget*: 1796–1797.
———. 1934. *Experiment in Autobiography: Discoveries and Conclusions of a Very Ordinary Brain (Since 1866)*. New York: The Macmillan Company.
Wilson, E.O. 1975. *Sociobiology: The New Synthesis*. Cambridge: Belknap Press.
West, Geoffrey. 2017. *Scale: The Universal Laws of Growth, Innovation, Sustainability, and the Pace of Life in Organisms, Cities, Economies, and Companies*. New York: Penguin Press.
Zuckerman, Phil. 2007. "Atheism: Contemporary Numbers and Patterns." *Cambridge Companion to Atheism* (Cambridge Companions to Philosophy), edited by Michael Martin, 47–66. Cambridge: Cambridge University Press.

Index

E. Jonsson, *The Early Evolutionary Imagination*, Cognitive Studies in Literature and Performance, https://doi.org/10.1007/978-3-030-82738-0

N

The manufacturer's authorised representative in the EU is Springer Nature Customer Service Centre GmbH, Europaplatz 3, 69115 Heidelberg, Germany. If you have any concerns regarding our products, please contact ProductSafety@springernature.com

Printed and bound by CPI Group (UK) Ltd, Croydon, CR0 4YY

30/09/2026

02225702-0013